AF560383

Vocational Guide to Servicing TV, Satellite and Video Equipment

Vocational Guide to Servicing TV, Satellite and Video Equipment

Rajneesh Goel

RANDOM PUBLICATIONS
NEW DELHI (INDIA)

Vocational Guide to Servicing TV, Satellite and Video Equipment

---

ISBN 978-93-5111-407-9
© Reserved

All Rights Reserved. No Part of this book may be reproduced in any manner without written permission.

Published in 2014 in India by

RANDOM PUBLICATIONS

4376-A/4B, Gali Murari Lal, Ansari Road
New Delhi-110 002
Phone : +91-11-43580356, +91-11-23289044
e-mail: randomexports@gmail.com, sales@randompublications.com, info@randompublications.com

Type Setting by : Keystoneprintads, Delhi-110051
Digitally Printed at: Replika Press Pvt. Ltd.

Preface

A genuinely practical hands-on guide for service engineers on the digital equipment. The cable has an advantage for subscribers who don't want digital programming because there's no equipment needed other than a television.

Channel surfers, in particular, might want to know whether cable, satellite, or Internet services offer the most viewing selections for the buck. But are the more channels really the merrier? "Cable services typically have certain tiers of channels," explains Salway, who also covers the broadband industry for about.com, an information and resource website. "They maybe start with 100 channels and then go up to a 150 channels. With satellite, you have the same type of choices, but you are stuck with a lot of situations you'll never watch." Having a selection of tiers - or channel packages with a fixed number of channels - might work for some viewers. But if you're not too keen on paying for channels you won't watch, the Internet TV service might be the answer to your prayers.

A satellite is basically any object that revolves around a planet in a circular or elliptical path. The moon is Earth's original, natural satellite, and there are many man-made (artificial) satellites, usually closer to Earth. A satellite is a smaller object in space which orbits around a larger object in space. Satellites can be either artificial, like the communications and weather satellites that orbit the Earth, or they can be natural, like our Moon.

A video camera is a camera used for electronic motion picture acquisition, initially developed by the television industry but now common in other applications as well. The earliest video cameras were those of John Logie Baird, based on the electromechanical Nipkow disk and used by the BBC in experimental broadcasts through the 1930s. All-electronic designs based on the cathode ray tube, such as Vladimir Zworykin's Iconoscope and Philo T. Farnsworth's Image dissector, supplanted the Baird system by the 1940s and remained in wide use until the 1980s, when cameras based on solid-state image sensors such as CCDs (and later CMOS active pixel sensors) eliminated common problems with tube technologies such as image burn-in and made digital video workflow practical. Set up, or set up and operate audio and video equipment including microphones, sound speakers, video screens, projectors, video monitors, recording equipment, connecting wires and cables, sound and mixing boards, and related electronic equipment for concerts, sports events, meetings and conventions, presentations, and news conferences. May also set up and operate associated spotlights and other custom lighting systems.

This book presents readers with an overview of the impact of TV, satellite and video on social participation.

I thank all members of my team who have helped in the preparation of the book. My special thanks go to "Random Publications" who have published the book.

– Rajneesh Goel

Contents

1

Television Receiving Equipment

DIFFERENT REQUIREMENTS FOR DIFFERENT VISUAL NEEDS

The techniques adopted to promote accessibility will need to cater to the needs of blind and partially sighted people. Blind people or people with very limited remaining vision may benefit from a text-to-speech facility, whereas those with a degree of usable remaining vision may be able to take advantage of adaptable interface that allows them to enhance the colour contrast between text and background or even zoom into the characters on the screen.

TERMS AND DEFINITIONS

Some of the definitions given below are tailored to the context of accessibility to television.

- Accessibility: Accessibility can be defined as a degree to which a person with disability is able to perceive, understand, navigate and interact with a service or a product. In the context of this report, it specifically focuses on the ability of blind and partially sighted people to use different television platforms currently available.
- Analogue television: Preceding digital television [DTV], all televisions encoded pictures as an analogue signal by varying signal voltage and radio frequencies. All systems preceding DTV can be considered analogue.
- Audio description: Audio description is like a narrator telling a story. An additional commentary describes body language, expressions and movements, making the story clear through sound. It describes what might otherwise be missed by a blind or partially sighted person. Audio description is also known as video description in some countries. See also Descriptive Video Service (DVS).
- Barrier: Any impediment, hindrance or obstacle that limits or

prevents the decorous, convenient and safe access, use, enjoyment or interaction with the environment.

- Captioning: Another term for subtitling used in many countries such as USA, Canada, and Australia. Even though the terms caption and subtitle have similar definitions, captions commonly refer to on-screen text specifically designed for hearing impaired viewers, while subtitles are straight transcriptions or translations of the dialogue. Captions are usually positioned below the person who is speaking and they include descriptions of sounds (such as gunshots or closing doors) and music. Closed captions are not visible until the viewer activates them. Open captions are always visible, such as subtitles on foreign language content.
- Descriptive Video Service®: Descriptive Video Service® (DVS) is the registered trademark WGBH created for its video description service; video description is another name for audio description and is widely used in North America. See also audio description.
- Digital Television [DTV]: DTV is the transmission of audio and video by digital signals, in contrast to the analogue signals used by analogue television. It supports the transmission of many more channels and also access services such as audio description and subtitles. DTV is gradually replacing analogue television and several countries such as Germany, Spain, USA and the UK have switched, or are in the process of, carrying out nationwide digital switchovers.
- Digital Terrestrial Television [DTT]: DTT uses the terrestrial aerial/ antenna signal, instead of cable or satellite, to broadcast.
- Digital Video Recorder [DVR]: Digital video recorder refers to set top boxes that have the capacity to record video in a digital format to a disk drive or other mass storage device.
- Internet television: Internet television is a streaming service distributed via the internet. It allows the users to choose programmes that they want to watch from an archive of programmes or from a channel directory. Some internet television services that have gained popularity in the past few years are:
  - RTÉ Player in the Republic Of Ireland,
  - Hulu and Revision3 in the United States,
  - Nederland 24 in the Netherlands,
  - ABC iView and Australia Live TV in Australia,
  - SeeSaw, BBC iPlayer, 4oD, ITV Player and Demand Five in the UK
- Internet Protocol Television [IPTV]: IPTV services can be grouped into 3 main categories:
  - Live television, with or without interactivity linked to the television programme in progress;

  - Time-shifted programming and catch-up television;
  - Video on demand (VoD): browse a catalogue of videos
- *Subtitling:* Communication support service that translates the oral dialogue and sound effects in any audiovisual production to text and graphics displayed on the screen. See also captioning.
- Subtitles for deaf and hard of hearing [SDH]: This term is often used to differentiate subtitles used to translate foreign language content from subtitles specially targeted at deaf and hard of hearing people.
- Synthetic speech: Synthetic speech is artificial human speech, which is produced by a computer. There are a number of different software applications through which this process can be achieved.
- Text to Speech [TTS]: Some speech synthesisers use pre-recorded human speech and fit words together to form sentences (this is most often used in applications with a limited vocabulary, such as a talking clock). Other synthesisers are more complex in that they fit together tiny portions of speech (sounds) to form words and sentences. Using this method, a synthesiser is able to produce an unlimited vocabulary, and can therefore read aloud any text input. This is known as text-to-speech synthesis.
- Universal design: Design of products and environments to be usable by all people, to the greatest extent possible, without the need for adaptation or specialized design.

PUBLIC SERVICE TELEVISION IN TRANSITION

The four countries studied here, namely the Czech Republic, Hungary, Poland and Romania differ in their pre-transformation broadcasting histories. The media policies (or, to use the term of the era, 'information policies') of the late communist party-states ranged from the traditional Soviet-type communist model where the media were completely subordinated to the political elite (Romania) to a more liberal model where journalists had considerable autonomy (Hungary), Czechoslovakia lay somewhat closer to the Romanian model and Poland to the Hungarian.

In spite of these historical differences, however, all of these countries encountered largely identical tasks in reforming their public media. All four countries have left the former East Central European broadcasting organization OIRT and joined the Western European public service association EBU.

The adoption of the Western European model of public service television included at least four different tasks:

- *Mission:* media policy makers were to make public service television introduce a program policy to suit society's needs in the emerging dual (i.e. public service and commercial) media landscapes.

However, there has not been any consensus on what this program policy should be.

- *Funding:* media policy makers were to establish a sound financial basis for public service television. At the same time, there was a general economic crisis in these countries during the transition years and, to a varying extent, throughout the 1990s. In addition to this, the low number of subscription fee payers and the small size of the advertising market have made the funding of public service television an acute problem.
- *Professionalization:* media policy makers were to promote the adoption of professional standards and, as a result, of a journalistic performance to comply with the needs of public service television. This was to be done at a time when the journalistic community was polarized and politically divided and, as a result, professional considerations were commonly subordinated to political ones.
- *Political independence:* media policy makers were to liberate public service television from political pressure. Authoritarian concepts of the media survived the political transformation. The dividing line between legitimate political marketing and illegitimate political pressure on the media had not crystallized yet. Many politicians in the region believed that their mandate entitled them to take control of the media. It was commonly argued that the journalists should support the democratically elected governments in the building of democracy rather than criticize them.

In this paper, I will compare the transformation of public service television in the above-mentioned four East Central European countries. I will focus on how the issues of mission, funding, professionalization and political independence are interrelated, and how they place public service television in some of the countries studied here into a vicious circle.

Mission

The European institutions, and in particular the Council of Europe have urged the countries of East Central Europe to deregulate the media, i.e. to terminate the monopoly of public television and to license private commercial broadcasters (Mungiu-Pippidi 2001: 76). Thus a major task for the broadcasting acts in the early 1990s was the setting up of dual broadcasting systems in the countries of the region. This was done in the following ways in the countries studied here:

In what was formerly Czechoslovakia, a federal broadcasting act was passed in November 1991. It remained effective after the division of the country under the title of the Czech Television Act, although some minor amendments have been made. In addition to the two channels of Czech Television, namely CT1 (news and mixed programming) and CT2 (cultural and minority programming), the law allowed for the operation of a third, private channel which started broadcasting in February 1994. TV Nova, the first commercial channel of national reach in East Central Europe, is run by Central European Media Enterprises and local investors. In 2000, CT1 and CT2 had a total of 32 percent audience share on average.

In Poland, the Broadcasting Act was passed in December 1992 and modified in 1995. There are three public service television channels, including two terrestrial and one satellite channel. TVP1 provides mixed programming, TVP2 focuses on culture and minorities, while TV Polonia addresses the Polish diaspora abroad. The first private channels that have been made possible by the Broadcasting Act, namely Polonia 1 (owned by the Italian Nicola Grauso) and POLSAT (owned by an emigrant Pole, Zygmund Solorz) started broadcasting in 1994. In addition to these, there are further private channels most of which broadcast through satellite or quasi-national networks of regional broadcasters. In 2000, TVP1 and TVP2 were watched by 48 percent of the national audiences on average, a share comparable with that in Western European democracies. Its success with the audiences is, however, at least partly explained by the commercialization of some of its programs.

In Romania, the Audiovisual Act was passed in the spring of 1992 and the Law on Public Service Radio and Television was passed in 1994 and modified in 1998. There are three public service television channels there including two terrestrial and one satellite channel: TVR1 provides mixed programming, TVR2 focuses on culture, while Romania International addresses the Romanian diaspora abroad. The biggest commercial channels include Pro TV (SBS Broadcasting), Antena1 (various foreign investors), and Prima TV (SBS Broadcasting). According to AGB Data Research, in 2000 TVR1 and TVR2 were watched by some 40 percent of the audiences, but with an uneven audience share with 37.7 percent watching TVR1 and only 2.5 percent TVR2. The third public service channel, the satellite-based Romania International, has hardly any domestic viewers. The relatively high audience share of public service television is also explained by the fact that rival commercial channels are unavailable in some of the rural areas.

Hungary's case differs from that of other countries in the region. No broadcasting act was passed in Hungary until the mid-1990s. The first years of democratic consolidation were characterized by a 'media war' for control of media policy between the right/conservative and the left/liberal political elites. Whereas the former insisted that the public service media must express national interests (which, they said, were identical with their own policies), the latter argued that the state must pursue a value-neutral media policy, and let the audiences decide (Bajomi-Lázár 2001: 85–94). The reason why no law was passed for years is that—unlike in other countries in East Central Europe—Hungary's broadcasting act was to have a two thirds majority. Finally, the radio and television was passed with a national consensus in

December 1995. Hungary has one terrestrial and two satellite public service channels: m1 (terrestrial) provides news and mixed programming, whereas m2 (satellite) broadcasts cultural and minority programs. The third, satellite public service channel, Duna Television, provides programs for Hungarian minorities abroad and was established in December 1992. After years of intense debate and struggle for the control of the then monopolistic public media, a compromise-based Radio and Television Act was passed in December 1995 that allowed for the establishment of private channels. The terrestrial private commercial channels RTL Klub (owned by CLT-UFA) and tv2 (owned by MTM-SBS) started broadcasting in October 1997. In 2000, m1 was watched by 12 percent of the audiences on average, whereas the two other public service channels were hardly watched at all, although Duna might have had a considerable audience share among the Hungarian national minorities in neighboring countries. The m1 channel's share of the 18–49 age group audience (i.e. the commercially most valuable age cohort) was less than eight percent (Jakus 2000/2001: 622).

All the media laws in the region define the mission of public service broadcasting in much the same way and refer to such concepts as impartial and objective news reporting, national culture, non-discrimination between the different sections of the population, as well as the representation of all views (Sparks & Reading 1998: 148–151; Mungiu-Pippidi 2001: 85). For example, Polish Television's charter, last rewritten in 1998, uses such concepts as "political neutrality", as well as "non-commercial" and "quality programming".

In comparison with other audiovisual laws, the Polish Broadcasting Act is unique in that it rules that Christian values need to be prioritized (Ociepka 2001: 115). Czech Television has to broadcast, among other things, "objective, open, generally balanced and all-embracing information" as well as a "balanced range of programs" (Czech Television Yearbook 2000: 90). Hungary's Radio and Television Act stipulates that public service programs include artworks belonging to universal and Hungarian culture, educational, scientific and religious programs, children's shows, minority programs, environmental broadcasts and, lastly, political news.

In general, the definitions of public service broadcasting in East Central Europe reflect the impact as the orthodox school. However, there is an important difference compared with the traditional concept of public service broadcasting: unlike the BBC, public service television channels in East Central Europe are allowed to broadcast advertisements. Evidently, advertising is not just a matter of funding; it has a huge impact on programming policies. Advertisements pre-structure programs and incite the broadcaster to provide entertainment, rather than traditional public service programs.

The broadcasting acts in the region are becoming outdated. They were conceived before commercial channels started operation and their program policy became known, before the internet became an everyday means of (mass) communication, before digital technology emerged, and before satellite and cable transmission became widely available in East Central Europe. The rise of commercial broadcasting has challenged public service television in various ways. Commercial broadcasters have professionalized television programming. They have introduced new, expensive production technologies, modern audiovisual solutions, dynamic media personalities and interactive shows. The large, bureaucratic institutions of public service television were usually slow to react to changes in television trends. As a result, commercial television has changed traditional viewers' habits: audiences have turned away from public service television. In particular, younger and more commercially promising viewers have switched to the commercial channels, whereas the older generations with lower incomes have continued to watch public service television. Thus commercial television has undermined both the legitimacy of public service television as a public institution and reduced its advertising revenues.

Apart from the rise of commercial television, the emergence of new technologies has imposed another challenge on public service television. Today, media markets offer a multi-channel supply. Some specialized, multi-language channels such as National Geographic and Spectrum for example, cover scientific and environmental issues 24 hours a day. Traditionally, educational programs have been considered a key reason for public service television to be maintained and funded from public resources, but why should the audiences pay the subscription fee if some commercial channels provide the same content free of charge? Besides, in the near future interactive technologies will allow for the individual selection of programs. Once users watch channels specifically designed for their tastes, or make their own selection through the Internet, mixed programming makes little sense.

In short, the question needs to be asked as to whether television continues to be the most efficient means of delivering traditional public service content to consumers. Some analysts argue that the different content traditionally associated with public service broadcasting should be delivered to users through different channels and especially via the Internet (Ildikó Kováts in BBC vagy RAI? 2001: 108–109). The counter-argument is that public service television has an integrative role in that it creates a forum for public deliberation in which people with different interests and tastes can participate simultaneously. The decline of such a public forum would result in the decline of the (political) community (Vajda 1999: 60).

What the role of public service television shall be is evidently an open question. It remains, however, that the current broadcasting acts in East Central Europe need to be redrafted in the near future in order to better comply with the changing technological environment and the subsequent changes in viewers' habits. The first generation of broadcasting acts in the region has opened the way for private commercial channels. It is the job of the second generation of broadcasting acts to redefine the role of public service television in this new context.

Funding

Funding has always been a crucial issue. Public service programs involve higher production costs than commercial programs do: whereas public service television channels are required to offer domestically produced shows such as documentaries and news, commercial channels are able to broadcast cheap imported products such as soap operas in most of their airtime.

In addition to this, public service television channels continue to be, or were until recently, vast bureaucratic institutions. Polish Television still has as many as 6,705 employees (Guide to Polish Television 2000: 58). Czech Television has 2,800 employees, which is, however, a 29 percent reduction in their numbers compared to previous years (Czech Television Yearbook 2000). Hungarian Television and Romania Television have dismissed nearly one third of their employees in recent years. For the sake of comparison, national commercial television channels employ 300–400 people (although they are not bound to produce as many home-made programs as public service channels are). Within the huge institutions of public service television, financial responsibility tends to be blurred. The costs of program production are not an issue and there is no internal competition among the different studios (this is not the case within the BBC, cf. Rácz 2001: 121–122). Filming in neighboring countries, which is quite common among television channels such as Duna Television, TV Polonia and Romania International that provide programming for the diaspora, involves further costs.

As already mentioned, all public service television channels in the region are entitled to broadcast commercial advertisements, although while public service broadcasting involves higher production costs than commercial broadcasting, public service television channels are usually more restricted in their advertising—for example, there is a limit on the amount of advertisement time compared with commercial television (Czech Republic). Similarly, they are either not allowed to interrupt films with advertisements (Hungary), or are unwilling to do so for fear of loosing their reputation as a public institution (Poland). This disadvantages them against their commercial competitors in the advertising market.

With the exception of Poland, the countries described in this study have small audience markets to offer to advertisers. The Czech Republic has 4,185,000, Hungary 3,869,000, Poland 12,505,000 and Romania 7,782,000 households (Mungiu-Pippidi 2001: 87). None of these countries have rich or sufficiently numerous middle and upper classes to sustain several television channels through commercial advertising and, as mentioned above, the rise of commercial television in the mid-1990s further diminished the advertising revenues available for public service television. Apart from advertising revenues, public service television channels are also funded through subscription fees. However, fee evasion is high in all of these countries. In Hungary for example, only about 68 percent of all household pay the fee.

Because advertising revenues and subscription fees do not cover all production costs, it is not uncommon for public service television channels to be under-funded, which then results in a dual dependency. On the one hand, it increases the dependence on business companies whose advertisements they broadcast. Czech Television, for example, decided not to broadcast an investigative report about one of its major advertisers after the company in question made a telephone call to the channel (Školkay 2001: 121). On the other hand, in some of the countries they are required to apply for government support. The governments' regular financial support has been an acknowledged practice in Hungary where at the end of 2000, the government covered Hungarian Television's debt of HUF 20,000,000,000 (approximately USD 70,000,000); at the end of 2001, public service television has run into another financial crisis and is in need of approximately HUF 12,000,000,000 further government support. Romanian Television has also received subvention from the government.

By contrast, in the Czech Republic and Poland, public service television is financially independent from the government. Czech Television, although it did produce a slight loss in 2000, has not received any government funding, covering its losses from financial resources created from retained earnings (Czech Television Yearbook 2000: 72). Polish Television has even been able to make a profit (Guide to Polish Television 2000: 58). Public service television in Romania has recently been able to produce some profit as well, although until 1999 it produced steady losses.

In summary, the financial situation of public service television channels varies across the region. Polish Television has a sound financial basis; public service television channels in the Czech Republic and Romania are still in a vulnerable position, while the financial situation of Hungarian Television is in a critical state. Financial difficulties are a major threat to the political independence of public service television: once in financial crisis, they have to apply for government funding. However, governmental support has a price: as a rule, politicians expect positive coverage in the public service media in exchange for their financial support.

Professionalization

The modernization of public service television in East Central Europe includes professionalization, i.e. the adoption of journalistic practice to the needs of reformed public service television. This is needed because journalistic practice in the region differs sharply from that in the United Kingdom where the concept of public service broadcasting was conceived.

What is the general practice of journalism in the Czech Republic, Hungary, Poland and Romania? Above all, reportage and opinion are blurred. There is an undeniable lack of investigative journalism (Andrew Stroehlein quoted in Školkay 2001: 113). It is not generally the case that all parties concerned are interviewed; there is no follow-up to the first coverage of scandals (Vajda 2001: 157–159). Journalists rely almost exclusively on 'accredited' news resources, which allows politicians to control the news agenda (Vajda & Kaposi 2001: 30). The typical journalist in the region advocates an elitist concept of culture, which can hinder their dialogue with the audiences (Gross 1995: 212–213; Magyari 2000: 100–101). Journalists are engaged intellectuals rather than distanced observers. News items are selected to harmonize with the political sympathies of editors. Newspapers and broadcasters are considered opinion leaders rather than open forums where opposing views can be contrasted. Former journalists become politicians and vice versa (Bajomi-Lázár 2001: 94–111). Journalists are inclined to comment from the position certain political groups rather than report in an unbiased manner (Prevratil 1995: 168).

In short, advocacy journalism prevails over the objectivity doctrine. This journalistic practice is a continuation of pre-war traditions. Before World War II, the various newspapers and media were partisan and promoted different political ideologies. Similarly, most of the current press and media in these countries are best described as displaying an external, rather than internal, plurality of views.

There is no tradition of neutrally critical journalism in East Central Europe, instead partisan criticism prevails. In other words, journalists criticize what the political opponents of the party dear to them do, while at the same time they are loyal to politicians whose ideology they share. The division of the journalistic community along political lines is most apparent in Hungary and Poland.

In considering the problem of the traditions of partisan journalism and external pluralism from the perspective of public service broadcasting, at least three problems should be noted. Firstly, political bias has been a recurring issue in public debates and publications on journalism in recent years. Audiences appear to cultivate the objectivity doctrine, rather than partisan journalism. The public seems to expect internal pluralism and neutral criticism from journalists. Journalists are the public's trustees. It is their duty to abide by public expectations.

Secondly, even if the norms of external pluralism might create a flourishing and interesting private print press, the concept cannot be applied to public broadcasting. Public service television is based on universal subscription fees and should for that reason display internal pluralism, i.e. address all segments of the audience.

Thirdly, the division of the journalistic community ensuing from the concepts of external pluralism and partisan journalism may endanger their professional autonomy. The journalistic community is not bound together by a feeling of solidarity and has no common group identity. Journalists' political identities prevail over their professional identity. Their political division results in their being unable to protect their common interests, which makes them vulnerable when in conflict with the political elite—especially when it comes to public service television, which political elites, left and right alike, have tried to take control of.

There is currently a slow but sure change in the norms of political journalism in the region. A paradigm shift from what can be called the traditional continental European model of journalism to the Anglo-Saxon one is now underway. The process is slow because young journalists are continually being integrated into the professional structures dominated by 'old-school' senior editors. But it is sure, since there is a steady market pressure pushing for a journalism that is able to address audiences larger than the supporters of any given political party. The globalization of mass communication, marked, among other things, with the growing popularity of global news channels, is also likely to forward such a paradigm shift.

The adoption of the objectivity doctrine is, in my view, a positive development from the perspective of journalists. It helps foster a journalistic community with universal standards and can for this reason be understood as a process of professionalization. It meets public expectations and is therefore a key condition for the future success of public service television.

Political Independence

The BBC's political independence relies chiefly on unwritten traditions of non-interference by government. One of the key questions of the reform of public service broadcasting in East Central Europe is whether a tradition of independence can be 'created' by institutional means.

The common solution that media policy makers in East Central Europe have had recourse to was the establishment of boards of trustees to work as a buffer between the political elites and the public broadcasters. The members of these bodies are nominated on a political basis in all four countries. In the Czech Republic they are nominated by the Chamber of Deputies although the candidates are presented by various non-political organizations (Czech Television Yearbook 2000: 91). In Poland, the members of the board are appointed by the National Broadcasting Council which, in turn, is elected by the Seym (the lower house of Parliament), the Senate (the upper house) and the President of the Republic (Sawisz & Mikulowski-Pomorski 1995: 87; Jakubowicz 1995: 142–143). In Romania, the board is elected by parliament, the president of the republic, the government and also by journalistic organizations (Mungiu-Pippidi 2001: 81). In Hungary, in addition to parliament, various non-governmental organizations may also delegate members to the board (the powers of the members delegated by NGOs being more limited than the power of those delegated by the political parties). The argument for the imposition of supervisory bodies on public service television was that this structure would free public service television from political pressure by means of balancing governmental and oppositional influence. However, the result was that political conflicts over who controls the media were delegated from parliament to the council (Gellért Kis 1997: 65).

The past decade has brought many examples of government intervention into the public media in all four countries. Political conflicts culminated around the appointment and dismissals of the boards of trustees and the general directors of public service television. In Poland, Lech Walesa dismissed the entire National Broadcasting Council in 1996. In Romania, a crisis occurred in October 1998 when the different agents in charge of the nomination of the boards of trustees (namely the president, the parliament, the government and journalistic organizations) could not agree on who the members of the board should be. In Hungary, the appointment of an incomplete (i.e. government-only) board of trustees to Hungarian Television in February 1999 and to Duna Television in February 2000 provoked widespread protest among domestic opposition and the representatives of the European Union. In the Czech Republic, the dismissal of former General Director Dusan Chmelicek and the appointment of Jiri Hodac caused massive street demonstrations in December 2000.

One of the great illusions of the transformation of the media has had to be abandoned. Although the passing of the broadcasting acts was expected to create a 'rule of law' in the sector and to put an end to the decades-long tradition of 'manual direction', practice has shown that political pressure persists. The institutional reform of public service television has failed to free public service television from political pressure. The failure of the reform was encoded in the political traditions of the region. Political elites, left and right alike, were uninterested in the liberation of public service television (Klvana 2001: 10; Mungiu-Pippidi 2001: 79). As long as the party system is not consolidated, i.e. political elites come and go, politicians are not interested to work out long-term solutions. They can never be certain that they will stay in parliament for another electoral cycle. For this reason, they are not interested in a public television that is critical of all governments, which they could make good use of, once in opposition. Instead, they play double or quits, i.e. exert pressure on public television in order to make it loyal.

Authoritarian reflexes die hard. Politicians in the region advocate outdated concepts of the media. The soviet-type agitation and propaganda model, which considers journalists to be 'the Party's soldiers' and 'the architects of the soul' persisted with the political elites. Many politicians are still convinced that the media exerts a huge impact on public opinion and for that reason positive media coverage is the best tool to maximize votes—a view that both empirical research and practice have disproved. Many think the media legitimizes those in power by advocating their policies—a concept which is contrary to the democratic idea that the media legitimizes power holders by watching them.

In East Central Europe, it was commonly argued that the journalists were to support the democratically elected governments in the building of democracy rather than criticize them. This approach rejects the 'public watchdog' concept of journalism and relies on the so-called 'majority argument' according to which, journalists (and especially those in the public media) should be loyal to, rather than critical of the democratically elected government. As a conservative Hungarian journalist put it,

"[t]he most powerful argument for loyal journalism is that democracy is based on free elections... and for this reason the elected government represents the people's will, it is the trustee of the public will. It is, therefore, entitled to limit the power of the press (which power is not derived from general elections), and to create opportunities to have its voice heard and let the public know its policies and objectives (through the public service media). The loyal journalist accepts this principle and meets the function of the gatekeeper while keeping an eye on the government's interests; he or she reports on events from the government's perspective, and protects the government's position." (Varga 2001: 205 [my translation – P.B.L.]).

The majority argument has frequently been used to legitimize political pressure on public service television. Critics, of course, consider this argument to be contradictory to the liberal approach which considers the journalistic community an autonomous 'forth estate'. The concept of loyal journalism equates the particular interests of the government's electorate with the general interests of the whole of society.

In fact, from the democratic theories of the press and the theory of the 'division of powers' it follows that anyone but the government may have a say in the supervision of public service television, because it is a primary function of the press to watch those in office. Even the control of the public media by the opposition seems a better solution than current practice. There is, of course, little chance that, once in power, any political force would ever acknowledge this and modify the current broadcasting acts.

In addition to the majority argument, another argument used for the control of public service television is that it is still under the influence of former communist editors and that it is the job of the democratically elected governments to 'clean' the media. After World War II, there was either no professional training for journalists (e.g. Hungary), or journalistic education was subordinated to direct Party control (e.g. Romania). Political loyalty was more important than professional qualifications. As a result, the journalistic community was largely counter-selected (Splichal 1994: 66–72). Journalists have been considered to be 'communist collaborators', which has made them extremely vulnerable when in conflict with the new political elite.

Consequently they have no other choice than to be loyal to the new political elite if they wish to preserve their jobs in the politically controlled public media.

The approach describing journalists as 'communist collaborators' is mistaken, at least in some of the countries studied here. From the early 1980s onwards, and especially in Hungary and Poland, the party-state's control of the media underwent a gradual process of erosion. As a result, journalists offered a steadily more realistic coverage of the world (Jakubowicz 1995: 130). They questioned the status quo and broke former taboos. They introduced democratic ideals and were thus a means of political re-socialization (Sükösd 1997/98: 13–17). In short, a part of the journalistic community played a great role in the demise of the state socialist régimes rather than prolonging their survival. Although the subsequent governments have all attempted to take control of the national broadcasting organizations in all of the countries studied here, it would be misleading to suggest that the public media has ever been fully dependent on the political elites. Efforts to take full control of the public media have never succeeded and have generally provoked resistance. The intensity of this resistance has been shown by the events in Poland in the early 1990s described as a 'television war', and as a 'media war' in Hungary throughout the decade; the discourse of the age used military terms such as 'camps', 'fronts' and 'arms'. (Ociepka 2001; Bajomi-Lázár 2001).

BACKGROUND - ACCESSIBLE TELEVISION

MOVING TO DIGITAL TELEVISION - CHALLENGES AND OPPORTUNITIES

Some accessibility issues are the same for both analogue and digital television e.g. accessing printed instructions and user manuals, recognising buttons on the remote control and reading on-screen text. These issues have an impact on both the systems. However, the increased functionality and complexity of digital television introduces some new barriers. These may be due to differences in hardware, user interface or programme content.

Increased functionality and the massively expanded number of television channels available through digital make it more difficult to locate a particular channel or programme. With old analogue systems offering as few as four or five channels, finding out what was on and changing to the channel of choice was relatively straightforward. However, with hundreds of channels, it becomes necessary to remember many channel numbers or use some sort of navigation system, usually presented as an on-screen electronic programme guide (EPG).

The increased functionality of digital television brings more choices, but the need to memorise controls or sequences of actions presents a usability problem, particularly for blind people. Increased functionality also allows more information to be given to the user – programme information, setup menus, programme guides, parental controls, etc. This information is provided using text and graphics displayed on the screen which can be difficult or impossible to read for people who are blind or partially sighted. These problems can be solved by allowing users to change the size and colour of on-screen text and by providing text to speech output for blind users.

Although the increased capability of digital television introduces new accessibility problems, it also solves some problems.

Notably, it allows for the inclusion of user-selectable audio description with all television programmes, something which is not possible with most analogue systems.

AUDIO DESCRIPTION

Audio description is a verbal description of the visual scene in a television programme, spoken by a narrator during the pauses between dialogue. It is provided as an aid to understanding and enjoyment particularly, but not exclusively, for blind and partially sighted people.

It is delivered as an auxiliary sound channel and control signals that can be selected to override the normal programme sound when the audio description track or DVS® is detected. To avoid incompatibility and clipping, audio description signals need to be recorded with the same line up as the main audio channels [-18 db for the stereo or mono in the UK]. These criteria are often referred to as ancillary audio standards in technical specifications.

Audio description delivery requires broadcasters to produce and broadcast the description soundtrack and receiver manufacturers to enable support for description in their products. A way to achieve this is to provide a framework for delivery of access services such as audio description. This framework involves governments adopting and mandating legislative measures to ensure audio description is delivered.

In view of the need for description on television, in 1996 the UK introduced the Broadcasting Act which made it mandatory for digital terrestrial programme services to provide description on at least 10% of their programming. The Communications Act (2003) extended this mandate to include cable and satellite services as well, so broadcasters and service providers were left with no choice but to provide description as an additional track that could be selected. In contrast to the model above, in 2008 Canada launched the world's first open description channel, The Accessible Channel, a 24 hour national, English-language, described video, closed-captioned, basic HD digital television [HDTV] specialty service.

In October 2010, the 21st Century Communications and Video Accessibility Act was turned into a law in the US. The legislation gives individuals with vision or hearing loss improved access to television programming, smart phones, the internet, menus on DVD players, programme guides on cable television and more. Many other countries such as Germany, France, Spain and South Korea have also had audio description on their television programs for some time now.

Delivery of Audio Description on Television

Availability of audio description on television is quite varied across the world. The technology required to deliver and render the basic description track is rather simple, being an additional component of the common audio-processing solutions. Audio-visual equipment that supports multiple audio channels are ideal for delivery of description, and most products dealing with broadcast television content are capable of providing adequate support for audio description. Audio description has been available on television since the 1990s. However, with the arrival of DTV, the technology used to deliver audio description has undergone constant evolution and it continues to progress with the advancements in technology.

ANALOGUE TELEVISION - SECONDARY AUDIO PROGRAMMING TRACK

Secondary audio programming (SAP) is a supplementary audio channel for analogue television that can be broadcast or transmitted by any transmission system including cable, satellite and IPTV. It was often used for an alternate language, or for the DVS® offered in the U.S before the digital switchover. The description track was combined with the original sound track on the SAP channel of televised programming. Analogue television systems used in most countries have not had the capability to include user-selectable audio description in this way.

Digital Television - Broadcast Mix and Receiver Mix

Today, the following mature accessibility services can be broadcast and received with regular DTV equipment:

Broadcast mix: An additional audio track consisting of the original audio and the narrator is pre-mixed at the broadcaster side and is transmitted in dual channel mode together with the original audio track in the audio elementary stream. Receiver mix: As an alternative to broadcast mix, the mixing takes place inside the receiver. The audio description sound track is received along with the main audio soundtrack and the digital receiver mixes them together.

Receiver mix offers certain advantages for the user which can be incorporated into the system, including the ability to adjust the sound level of the description track, and routing the description to headphones so that only one person can hear it while others in the room hear the regular audio track. Both of the above mixing methods provide what is, in effect, 'closed' audio description. This means that the audio description is separated from the main programme audio in a way that the individual viewer can choose whether to hear it or not. This contrasts with 'open' description which is mixed in with the main audio so that all viewers receive it without having a choice.

In the absence of the delivery platforms as mentioned above, 'open' description also exists i.e. The Accessible Channel [TAC] in Canada broadcasts open description on 100 per cent of its programming. The channel broadcasts programmes that have been quite successful in the past but this time around, they are broadcast with AD. About 75-80% of the programmes on the channel have never been aired with audio description before (TAC, 2010).

SPOKEN OUTPUT FOR INTERFACES

As highlighted in the report 'Developer's Guide to Creating Talking Menus for Set-top Boxes and DVDs [National Center on Accessible Media (NCAM, 2009)] the highly visual nature of interfaces used in new digital media formats has created serious and growing barriers for blind and low-vision consumers. The more visual the interface, the harder it is for a blind user to use it. The provision of an audio interface for people with visual impairments will resolve issues relating to use of the visual interface.

In August 2010, INTECO, the Spanish National Institute of Communication Technologies, released a DTV operating system that features similar spoken output and control of display settings such as the size and colour of text in the menus and EPG. This operating system consists of an open source software solution that can be integrated into any digital set top box or integrated television. Smart Talk Freeview digital box, the first commercially available terrestrial set top box with spoken output of menus and EPG information went on sale in the UK in 2010. The set top box features a fully talking Electronic Programme Guide (EPG), spoken output of all menu settings and one-click access to audio description through a dedicated button on the remote control.

In addition, "Speech Solutions for Next-Generation Media Centers" (NCAM, 2009) developed open source software for talking set-top boxes built on the Linux-based MythTV platform. Text to speech [TTS] provision opens many doors for people with sight loss as an alternate means of navigation in this graphic-rich environment and enables independent access to new programmes and services on the television. It also demonstrates that TTS technology for television is technologically possible. Where spoken output is not provided, the thoughtful choice of tones or audible feedback can assist greatly in low vision or non-visual use.

Audible feedback- characteristics

Audible feedback is used to acknowledge a command given by the user, possibly for an operation, potentially a person with sight loss, uses a television receiver e.g. audible confirmation that a key has been pressed on the remote control. It must be noted here that the feedback has the potential to indicate a number of operations:

- Feedback announcing the receipt of a command from the user or the start of an action
- Feedback prompting the end of an action
- Feedback confirming the receipt of a standard command or in another case the receipt of an unrecognised command
- It is critical that this feedback is unique in its temporal pattern so that it can:
  - Be understood without giving further instruction to the user and
  - Not be confused easily with other audible feedback signals used in the same product or those in another product used simultaneously and in the same place.

Since audible feedback is less tangible than spoken commands, there is a definite need for the temporal pattern of different feedback signals to be distinct so that the user is able to differentiate one signal from the other i.e. {onomatopoeic description of different signals} Pip, Peep, Pip-pip [in quick succession], Pi.pi.pi.peep (slowly), Pip, pip, pip, pip, ...(specified times, slowly).

Often a user is operating a toggle, where a feature is switched on and off by use of the same remote control button. Here, the use of ascending and descending pairs of tones are commonly used to indicate on and off respectively. There are various ways in which blind and partially sighted people access information. For example, some use large or modified print materials, some use Braille and some use audio information. Audio information can broadly be split into two types-Spoken output, which comprises information that is read out by another person - either live, or on a recording or via synthetic speech. Audible feedback in the form of auditory signals e.g. Pip, Peep, Pip-pip [in quick succession]. These are normally used to confirm the receipt of a signal or completion of a task.

ON-SCREEN DISPLAYS

Adaptable fonts, changeable colours, simple uncluttered layout with zoom functions are all attributes that can transform a completely inaccessible on-screen display into a more usable product for blind and partially sighted viewers. However, lack of demand in the mainstream market has so far prevented equipment manufacturers looking into or providing features that will support the needs of blind and partially sighted people. These functional attributes have been discussed in section 4 on user requirements.

REMOTE CONTROLS

The remote control is the principal piece of equipment that is used to interact with the television set, so due consideration needs to be given to its design and functionality. DTV, equipment and services are more or less unusable without a remote control. Several studies conducted around the accessibility of remote controls for people with disabilities have published a set of recommendations that relate to its ergonomics and utility. For example, the guide for remote controls in Handbook of Adult Anthropometric and Strength Measurements includes suggestions for its nomenclature and button size/positioning. These recommendations have been discussed in the section 4.2.2.

INTERNET TELEVISION AND IPTV

Internet television and IPTV are relatively new platforms for viewing television content that are fast gaining popularity and due to improvements in internet speeds, they are likely to become more and more popular. Internet television (sometimes known as online television) is a television service distributed via the internet. It allows the viewers to choose programmes that they want to watch from an archive of programs or from a channel directory.

Internet Protocol television is a system through which television services are delivered using the internet and broadband internet access networks, instead of being delivered through traditional radio frequency broadcast or satellite or cable television. Internet Protocol Television: IPTV services can be grouped into 3 main categories:

- Live television, with or without interactivity linked to the television programme in progress;
- Time-shifted programming and catch-up television;
- Video on demand (VOD): browse a catalogue of videos

Watching content over Internet Protocol Television [IPTV] involves watching content on a system that has a browser and an internet connection. It is advisable to use User Interface Web Browser Guidelines where content is delivered through a web browser. These guidelines identify characteristics that would provide a more adaptable user interface to allow users, not only those with disabilities, to have a much greater selection in what suits them most. These guidelines can be found on http://www.w3.org/WAI/UA/wai-browser-gl#Introduction to WWW Browser.

However it is the Web Content Accessibility Guidelines [WCAG] that will be applicable for content that is made available on the internet. Web "content" generally refers to the information on a web page or web application, including text, images, forms, sounds, and such. Since, both technologies deliver television, and because technology varies across the world, this document covers both.

CONSUMER EQUIPMENT STANDARDS

The underlying aim of this document is to encourage all relevant stakeholders to make access services more widely available on television and as a result focuses on the significance of standardisation across the globe and technologies. This is deemed absolutely essential for interoperability and usability of access services on television and therefore impacts on product and service design. A comprehensive list of standards available in this area of operation can be found in appendix 3 of this document. These standards apply to, amongst a host of other functions, access services, design of remote controls, user interface, and symbols that aid accessibility. All such design should conform to relevant consumer equipment standards whilst also not being overly prescriptive to avoid stifling innovation.

ACCESS TO TELEVISION - USER REQUIREMENTS

This section on user requirements identifies and details the requirements which, if implemented, will facilitate access to television for blind and partially sighted people. These requirements are applicable regardless of television viewing platform and geographical location. They will not only improve access but will make television viewing a more satisfying experience for the target audience.

The user requirements have been assigned to one of the following three subdivisions:

- Must have: this means that the requirements are necessary
- Should have: this means that the requirements are recommended
- May have: this means that the requirement could add positively to the experience.

Current Barriers - Access to Television

This section is divided into two parts -

- Barriers for blind people and
- Barriers for partially sighted people.

It is important that stakeholders dealing with accessibility of television are aware of these barriers while designing new products.

People who are blind:

- May struggle with complex packaging;
- May not be able to see (to read);
- May not be able to see what is displayed on visual display units;
- May not be able to see visual feedback of operation i.e. connector cable initial set up successful;
- May not be able to access information presented (only) via graphics or text;
- May not be able to fully understand and enjoy television programmes if good quality audio description is not available.

People who are partially sighted (low vision):

- May have difficulty discriminating text from the background if contrast is insufficient;

- May have difficulty discriminating colours;
- May have difficulty with glare – from environment or the surface of the object;
- May not see (to read) signs, labels and text:
  - If text is too small for them,
  - If contrast with background is too low,
  - If text is presented as small raised letters (same colour as background),
  - If information is coded with colour only (colour deficiency),
  - If there is glare if they have light sensitivity,
  - If there is insufficient ambient light.
- May not be able to read moving or scrolling text

Packaging and Getting Started

- Packaging must be easy to open.
- Packaging must not contain materials that may cause injury, such as staples;
- The packaging must convey clear information to the buyer about the functions of the equipment. Where appropriate, official recognised logos must be used. Tactile alternatives must be provided when possible.
- Additional components such as batteries and other accessories should be separately and securely wrapped but easy to open by hand.
- Accessible print standards, as applicable in different countries, may be used for any text on the packaging such as RNIB Clear Print Guidelines for the UK.

Instruction Manual Including Quick Start Guide and Main user Guide

- Full user guide and the quick start guide must be available, on request, in alternative formats – Braille, Large Print, Audio and accessible online formats.
- Information presented using diagrams and screen shots must also be available within the text instructions.
- The instructions must provide information about the accessibility features of the product and how to access audio description.
- The design and layout should conform to the guidelines/ standards recommended by blindness organisations in different countries.
- Glossy paper should be avoided.
- Page layout should be simple and uncluttered.
- On-screen information may also be provided, in addition to the printed manual.

Identifying Connectivity Options

- External connections must be easily accessible and clearly marked.
- If information about connector engagement is displayed on the screen then, clear audio feedback of connector engagement must be available. i.e. Scart Cable or HDMI installed

Equipment Tuning

For example: initial settings, on-screen display, basic tuning and advanced settings for access features such as subtitling/ audio description/ favourites.

- Receivers must carry out the full tuning sequence automatically following initial powering on once the receiver is connected to a display, a power source, or service platform [terrestrial, cable, satellite].
- All receivers must automatically identify new services launched by broadcasters without the user having to retune.
- At all stages of the tuning sequence the status should be indicated in text and audio, describing the action being carried out, state of progress and time remaining.
- This screen should also have a clear 'exit' or 'skip' prompt.
- Following initial tuning, an on-screen message should prompt users of the options to set up any preferences. This may be achieved by referring them to the Quick Start Guide or by an on-screen step-by-step process. Typical preferences essential to this set of user requirements are:
- audio description default on
- set-up of favourites or 'hide' channels (as appropriate)
- power saving modes or timers
- text to speech default on

USER INTERFACE AND REMOTE CONTROL

- It is critical that the user interface is designed in such a way that it takes into account the needs of all possible users, including blind and partially sighted persons. Some of the key functions, inclusive but not limited to, that must be completely accessible are;
- switching on and off,
- changing channel,
- adjusting volume,
- accessing the EPG and
- be able to turn on/off access features such as audio description.

This section has been split into five parts for ease of navigation and understanding;

- Navigation

- Interactivity
- Remote control- design and functionality
- Audible feedback
- Speech synthesis or text-to-speech

Navigation

- Only the visible or the safe text area, where text will not be cut regardless of the over scan of the television, must be used.

Note: Safe area is a measured zone within the video frame which defines where all text should be contained to prevent loss during transmission and reproduction.

- Only one font should be used throughout the application.
- Changes in font size and colour should be minimized.
- Colours with a saturation index of less than 85% should be used to avoid distortion and flicker.
- The text colour should have sufficient contrast against the background colour.
- The font used should not distort and the size used shall be sufficient to assure the legibility of the text. Tiresias is the recommended font for this purpose in the UK, although others as recommended by organisations in different countries of the world can be used.
- Use of multiple columns for on-screen text may be avoided, as these can be difficult to read. If multiple columns are used, a sufficient margin may be provided between the columns with an appropriate column size when the selected font size is large.
- Users may be given the option to change colour combinations, font sizes and the screen background, including a "high contrast" display option, with larger text and icons, dark background and light text. These parameters may be configurable by the user.
- The text paragraphs may be kept short with good line spacing.
- The excessive use of graphics may be avoided to represent different options.
- Icons may be accompanied with text.

Interactivity

DTV receivers should be designed using universal design principles such that they do not create any access barrier for people with functional diversity. The user requirements in this section are applicable across the number of interactive services that area available via the television equipment these days such as internet television, IPTV, red button services, catch-up services.

[Must have]

- All interactive services must have the ability to deliver audio description.

- The service must let the users choose their preferred text size/ background and foreground colour for the on-screen display.
- The service must let the user enable TTS or audible feedback to hear an audio alternative to the displayed text.
- Information in the menus must be perceptive and logical. Menu structures must be simple, each following a similar pattern path.
- On-screen display must show information about programmes, such as cost of pay-per-view, terms and conditions, and availability of access features such as audio description. These must be available on all IPTV services through a single button.
- Information mentioned above must be available as text on-screen as well as in TTS.
- Information chosen by the user for display, such as the channel guide, genre guide, or information on television content, should remain on-screen until the users decides to remove it.
- Web Content Accessibility Guidelines (WCAG) 2.0 should be followed where appropriate to ensure complete access to the content on a web page.
- User Interface Web Browser Guidelines should apply where content is delivered through a web browser to allow a more adaptable user interface.
- In conjunction with the remote control there should be a single button that returns the viewer to the opening menu in IPTV.
- The user should be able to turn on and off TTS and audio description on a temporary or permanent basis.
- The user may be able to change the layout according to their preference, so that the layout becomes simpler and more intuitive.

REMOTE CONTROL- DESIGN AND FUNCTIONALITY

The remote control is, without a doubt, the principal piece of equipment used for interacting with various television applications such as EPG, mini guide and access features. However, little has been done to ensure that they are usable by persons with sight loss. To be rendered completely accessible, thought not only needs to be given to its design but also its functionality. The following user requirements recommend ways in which this can be achieved.

Remote Control Design

Remote controls must be ergonomic and usable by people with sight loss, in accordance with universal design principle

- Buttons must be logically and functionally distributed e.g., channel numbers to be grouped together, channel up and down beside the volume up and down. Consideration must be given to space between buttons with sufficient space between them. This must be done in accordance with guidance already specified in technical standards such the D-book 6.0, (DTI, 1998)

- Buttons must be of suitable size for use by people with sight loss. [Refer to D-book 6.0, DTI (1998)]
- The number 5 button must have a raised dot or line for tactile identification. [ES 201 381 Human Factors (HF) ETSI Standards, 1998]
- Where a purely touch sensitive remote control is provided with the equipment, another remote control must be made available upon request which is more accessible for people with sight loss. This is similar to the current strategy that has been adopted by Apple Inc. for their touch products.

[Should have]

- The On/Off button should be sufficiently isolated so that it cannot be accidentally pressed.

Remote Control Functionality

- The angles of transmission and reception between the remote control and the television must be broad so that it is not necessary to orient the control accurately. [Trace R & D Centre, University of Wisconsin-Madison (1998) Accessible Design of Consumer Products, Section 1: Output/Displays]
- The remote must provide one-touch access buttons to accessibility services i.e., audio description and subtitling.
- No action should require pressing two or more buttons at the same time.
- Assigning dual functionality to buttons should be avoided.
- The provision of basic operating controls on the receiver itself is encouraged. These should provide at least a minimum level of operation without use of a remote control. These should be labelled clearly and meaningfully using the same labelling as used on the remote control.
- Tactile feedback should be provided on button presses.

AUDIBLE FEEDBACK

Audible feedback is an important tool for communication, compensating any inability to process visual feedback or on-screen information by providing the user with adequate complementary information.

[Must have]

- Any process that takes place on-screen (e.g. system updating itself), in addition to being reported by a text to speech engine, must be accompanied by an audible feedback that allows the user to identify the process that is taking place.

- Pop-up messages must be accompanied by an audible feedback.
- The fundamental frequency of the signal must not be higher than 2.5 kHz
- When the channel is changed sound signals should identify television broadcasts that contain audio description for the blind and partially sighted users. This should also be indicated while scrolling down the EPG.
- In the case of an operation confirmation signal, if the user selects the next operation, the reaction of the next operation should be given priority and the former auditory signal should be interrupted.
- The user should be able to turn off insignificant auditory signals, by turning off the extra beeps. [e.g. "extra beeps on/off"]

Text-to-Speech [TTS]

Digital set-top boxes offer access to a wealth of information, entertainment and services via electronic program guides (EPGs), which require users to scroll through long lists of on-screen text and graphics to view choices and select a programme or service.

The latest generation of digital set-top boxes offer EPGs that provide detailed information about programmes, the ability to set parental controls, and the ability to programme channel selections for future viewing. However, since most of the information on the EPG is of graphical nature, it is completely inaccessible to a blind person and extremely hard to follow for a partially sighted person. The more graphical the interface, the harder it is for a user who is blind or partially sighted to use it.

TTS is automated verbal generation of elements presented on-screen [graphics, text, icons etc] in simulated human speech. A system used for this purpose is called a speech synthesiser and can be implemented in software or hardware. The user requirements for TTS are as follows:

- An audio prompt must be provided at start-up that instructs the user how to enable TTS features.
- The user must be able to activate or de-activate TTS as per need.
- A single button on the remote control must be assigned at start-up to toggle the enabled/ disabled setting.
- The user must be given the control to increase or decrease the volume of TTS relative to the broadcast level. A number of factors could influence the user's decision to change the volume e.g. hearing ability of the user, distance from the product, ambient sounds etc.
- The user should be able to change the level of verbosity depending on how much spoken cues is preferred.
- The user should be able to adjust the speed of the TTS audio and other characteristics such pitch and TTS voice type.
- The TTS should be available in different languages. Where there is

receiver UI language that is not supported by TTS, then the user must be informed in some way before changing the UI language.

- The information reproduced through TTS may skip unwanted redundant elements in the information. Please see "A Developer's Guide to Creating Talking Menus for Set-top Boxes and DVDs" for further details (NCAM, 2009).

ADDITIONAL SERVICES

Parental Lock

- Users must be able to lock access to certain channels depending on the content being broadcast on the channel. This is only in cases where this feature, parental lock, has been made available on the system.

Audio Description

- Television receivers must be able to deliver audio description in cases where content providers have made a description track available.
- There receiver must have the ability to enable audio description so that the setting remains active across channels.
- Events that carry subtitles and/or audio description must be clearly indicated in the EPG in a manner that the user can identify them without having to hunt for them. Appropriate and relevant logos must be used.
- Programmes that support audio description must additionally be identified by differing audible feedback signals (e.g. beep) when description mode has been globally selected and deselected and when a current described event is selected.

Recording and Playback on DVR

- Clear indication must appear on-screen accompanied by an audible signal [e.g. beep] when recording is initiated or stopped on a current event.
- DVR and On-demand services must also support recording and playback of audio description.
- When the hard disk of the DVR is almost full a warning message may be displayed on the screen whenever a recording starts or a new recording is programmed. This message may also be available in TTS.

Applications

- The programme name, channel number, event name and

information pertaining to the currently selected event must be voiced.

Emergency Information

- All emergency information that is displayed in text must be voiced.

Miscellaneous Information

- Where appropriate, the message must give instructions in text and speech on how to clear the message and how to save it for a later reminder.
- If the message requires user initiated or automatic re-configuration of the tuner system there should also be a warning in text and speech not to unplug the tuner while this action is taking place.

THE BEST TV EQUIPMENT AND SERVICES

The way we watch TV is changing, thanks to a nationwide transition to digital-only TV. The UK government is gradually switching off the current TV analogue broadcast signal. Switchover has already been completed in a number of regions and by 2013 all of the UK will have switched, meaning that you'll be able to watch and record television only if you have a digital TV service and appropriate equipment, such as a set-top box or a PVR.

THE BEST DIGITAL TV EQUIPMENT

If you're upgrading your TV equipment to digital, the best option for you depends on your budget and whether you want to record digital TV.

You'll need digital TV equipment for every TV set you want to watch after the switchover.

- Add a set-top box. A Freeview or Freesat set-top box is your cheapest digital option. Most TVs, even older 'box style' CRT sets, can be converted by connecting a set-top box. Prices start from as little as £40. Sky, Tiscali and Virgin Media provide their own set-top boxes as part of their digital TV subscriptions.
- Buy an LCD or Plasma TV with a digital receiver built in (IDTV). If you're in the market for a new television set, all the TVs we've tested have a Freeview receiver built in, so you won't need to add a Freeview set-top box. Some of the TVs we test have Freesat built in, too – use our product picker to find Which? Best Buy flat-screen TVs to suit all budgets.
- Use a personal video recorder (PVR). PVRs let you record digital TV programmes. Instead of tapes or discs they record programmes on to an internal hard disk. Unlike video recorders (VCRs) and many DVD recorders, most PVRs let you record one TV show while watching another (make sure the one you choose has two digital

tuners). PVRs also act as set-top boxes, so if you have a PVR you won't need a set-top box for the same TV.

Enhanced TV Features

In addition to offering more channels, digital television offers enhanced features. For example, most set-top boxes support an electronic programme guide (EPG), which is a regularly updated onscreen TV guide. By pressing the red button on your remote control, you can often find out more about the programme you're watching.

Sky and Virgin Media offer further interactive services, such as TV shopping, emails, games and banking. There are also some useful features for people who are hard of hearing or visually impaired. Audio description, for example, is a narrative that explains what's happening on the TV screen for people who can't see it clearly. Other services include talking EPGs, onscreen signing and recordable subtitles.

The Best Digital TV Service

The equipment you choose may depend on the digital TV service you want to watch. There are four main options: digital terrestrial television (Freeview), cable TV, satellite TV and online TV.

Digital Terrestrial Television (DTT)

DTT is better know as Freeview and lets you watch up to 50 digital TV channels. With Freeview, digital TV signals are received by the same TV aerial that you use to receive an analogue TV signal. Freeview comes built in to most new TVs; to convert an old TV you'll need to buy a separate Freeview set-top box. There are no ongoing subscription costs. If you live in an area with a particularly strong Freeview digital signal, you might be able to get away with a TV set-top aerial. Otherwise you'll need to ensure that your rooftop TV aerial is in good working order. Most people won't need a new aerial to be able to receive Freeview after the switch to digital (see will I need a new TV aerial?), or if you want to buy an indoor aerial, take a look at our reviews of Which? Best Buy indoor aerials.

Cable TV

This is only available on a subscription basis from cable company Virgin Media. The TV signal is carried via an underground cable, which may also double as a phone or broadband line. Cable TV is available to just over half of the UK population and you'll need a Virgin Media set-top box to connect to your TV. You can see how Virgin Media bundles measure up to rivals for customer satisfaction in the Which? review of phone, internet and TV packages.

Satellite TV

Satellite TV is available to most of the UK and there are two options for watching it. Both choices require you to have a satellite dish attached to your house.

SKY TV

Sky offers two services, both of which require a dedicated Sky TV set-top box:

- Sky subscription TV – where you pay a monthly fee to receive a wide range of television channels.
- For a one-off payment you can get Freesat from Sky, Sky's satellite TV alternative to Freeview.

Freesat from the BBC and ITV

The BBC and ITV have teamed up to offer another satellite alternative, also called Freesat. As with Sky's version of Freesat, you pay a single upfront fee with no ongoing TV subscription costs. Some new TVs have Freesat built in; otherwise you'll need a set-top box with Freesat capabilities. Some Freeview set-top boxes also have Freesat built in.

Online TV

BT offers a fourth option - TV over a broadband line - but you have to have broadband with BT to get its TV service. BT's TV over broadband service is called BT Vision. The Which? guide to cable TV, satellite TV and digital TV packages has more on BT's internet TV.

THE CHEAPEST DIGITAL TV OPTIONS

The cheapest ways to receive digital TV are with Freeview, Freesat from the BBC/ITV or Freesat from Sky. These require only a one-off payment for the digital TV equipment that lets you receive a basic range of digital TV channels. Basic Freeview set-top boxes start from £20, though you'll need to pay a bit more if you want a Freeview box that doubles as a PVR. Freesat boxes tend to be more expensive, though prices are falling. You'll need to pay a monthly fee if you want other digital TV options, including a Sky subscription TV deal or cable TV from Virgin Media.

TELEVISION AND RADIO EQUIPMENT

AUTOCUE

Autocue is a UK based manufacturer of teleprompter systems. The company was founded in 1955 and licensed its first teleprompter, based on a patent by Jess Oppenheimer, in 1962. The company began by producing teleprompting equipment, and now provides prompting, scripting, production and newsroom software applications and hardware. Autocue provides international prompter sales and rentals to countries outside of the Americas, while QTV handles prompter sales and rentals in North and South America.

HISTORY OF AUTOCUE AND TELEPROMPTING

Prompting Begins on Paper

Prompting began with Jess Oppenheimer, a writer, producer and director on the TV show "I Love Lucy" in the early 1950s. To solve the problem of the actors forgetting their lines, he developed a rudimentary prompting system. Oppenheimer took out a patent on the system. He licensed the patent to the teleprompting company Autocue in 1955. Meanwhile, a separate entity, QTV, was established in the US. Both companies started by renting teleprompting equipment to studios.

Oppenheimer's paper roll system survived until 1969 when Autocue introduced the first closed-circuit prompter. This used a closed-circuit camera system to film a scrolling paper script and display the image on a monitor attached to the front of the camera that was shooting the presenter. Use of a two-way mirror system allowed the script image to be reflected onto a sheet of glass in front of the camera lens, meaning that the presenters were able to read their lines straight from the script while looking directly into the camera. The mirror system meant that the image of the script was not visible to the main camera lens, and indeed this is still the way that most teleprompting systems operate. QTV followed shortly afterwards with similar technology. In the 1970s both companies started selling hardware in addition to maintaining their rental operations.

In 1984, QTV acquired Autocue and retained the two brands in their respective regions. The following year, the newly formed Autocue Group released a computer-driven prompting system, ScriptNet. At the same time the first newsroom computer systems were beginning to appear in television stations.

Prompting Enters the Digital Era

Autocue and QTV were also interested in digital prompting in more areas along with television news, leading to the development of its own scriptwriting and running orderpackage. This package used by many newsrooms, conferences, sitcoms, major drama productions, and more. This system worked on an early version of Microsoft Windows, and was named WinCue. In 1994, the Autocue Group created the first flat screen prompters.

Meanwhile, WinCue continued to accrue functions and features as customers (particularly TV newsrooms) demanded ever greater integration between the different elements of their workflow. Journalists need sources, and the ability to ingest, index and process stories from news agency wires was an early addition. Another requirement was the ability to control multiple playout devices, such as videotape machines and character generators, from a central running order. The Autocue Group accordingly developed its own automation system, as applicable to general programme playout as to news. This was followed by the ability to handle media as well as words within the production environment. Following a project with CNN, a wireless tablet-PC based system to effect and share script changes using 'digital ink' was added. These developments were not completely organic in origin. In 1998, a newsroom application company, DCM, based in Charlotte, North Carolina, was acquired to complement the existing newsroom product range. The Group retains an office in Charlotte to this day.

Networked and Integrated Teleprompting

Autocue have expanded into an integrated suite of tools that cover every aspect of broadcast production: planning, news gathering and ingest, production, scripting, running-order management, playout automation, media asset management and, of course, prompting.

Video Servers

In 2011 Autocue announced and launched its next generation video server, up to this point, Autocue ingest and playout modules had been discreet black boxes hidden in amongst Autocue's automation, scripting, media asset management and news tools. The latest version is a Linux based box designed to work as part of an Autocue or third party production workflow or as a standalone box in its own right. It is one of the most flexible solutions on the market offering a large number of support codecs and formats. At the end of 2011 over 100 channels of Autocue's latest ingest and playout product have been sold.

CATHODE RAY TUBE

The cathode ray tube (CRT) is a vacuum tube containing one or more electron guns (a source of electrons or electron emitter) and a fluorescent screen used to view images. It has a means to accelerate and deflect the electron beam(s) onto the screen to create the images. The images may represent electrical waveforms (oscilloscope), pictures (television,computer monitor), radar targets or others. CRTs have also been used as memory devices, in which case the visible light emitted from the fluoresecent material (if any) is not intended to have significant meaning to a visual observer (though the visible pattern on the tube face may cryptically represent the stored data).

The CRT uses an evacuated glass envelope which is large, deep (i.e. long from front screen face to rear end), fairly heavy, and relatively fragile. As a matter of safety, the face is typically made of thick lead glass so as to be highly shatter-resistant and to block most X-ray emissions, particularly if the CRT is used in a consumer product. CRTs have largely been superseded by newer display technologies such as LCD, plasma display, and OLED, which have lower manufacturing and distribution costs.

The vacuum level inside the tube is high vacuum on the order of 0.01 Pa to 133 nPa. In television sets and computer monitors, the entire front area of the tube is scanned repetitively and systematically in a fixed pattern called a raster. An image is produced by controlling the intensity of each of the three electron beams, one for each additive primary color (red, green, and blue) with a video signal as a reference. In all modern CRT monitors and televisions, the beams are bent by magnetic deflection, a varying magnetic field generated by coils and driven by electronic circuits around the neck of the tube, although electrostatic deflection is commonly used in oscilloscopes, a type of diagnostic instrument.

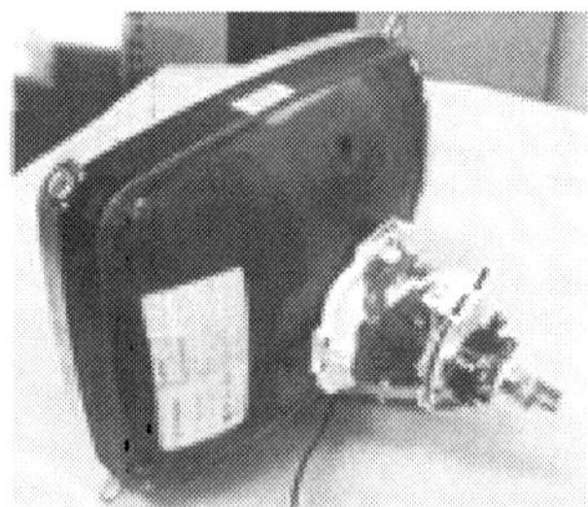

Fig. A 14 inch cathode ray tube showing its deflection coils and electron guns

Fig. Typical 1950s United States television set

Fig. Electron gun

HISTORY

The experimentation of cathode rays is largely accredited to J. J. Thomson, an English physicist who, in his three famous experiments, was able to deflect cathode rays, a fundamental function of the modern CRT. The earliest version of the CRT was invented by the German physicist Ferdinand Braun in 1897 and is also known as the Braun tube. It was a cold-cathode diode, a modification of the Crookes tube with aphosphor-coated screen.

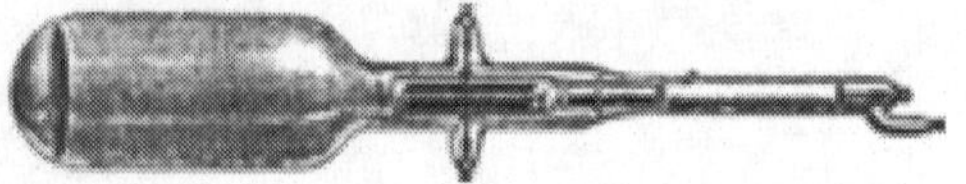

Fig. Braun's original cold cathode CRT, 1897

In 1907, Russian scientist Boris Rosing used a CRT in the receiving end of an experimental video signal to form a picture. He managed to display simple geometric shapes onto the screen, which marked the first time that CRT technology was used for what is now known as television. The first cathode ray tube to use a hot cathode was developed by John B. Johnson (who gave his name to the term Johnson noise) and Harry Weiner Weinhart of Western Electric, and became a commercial product in 1922. It was named by inventor Vladimir K. Zworykin in 1929. RCA was granted a trademark for the term (for its cathode ray tube) in 1932; it voluntarily released the term to the public domain in 1950. The first commercially made electronic television sets with cathode ray tubes were manufactured by Telefunken in Germany in 1934,

OSCILLOSCOPE CRTS

In oscilloscope CRTs, electrostatic deflection is used, rather than the magnetic deflection commonly used with television and other large CRTs. The beam is deflected horizontally by applying an electric field between a pair of plates to its left and right, and vertically by applying an electric field to plates above and below. Oscilloscopes use electrostatic rather than magnetic deflection because the inductive reactance of the magnetic coils would limit the frequency response of the instrument.

Phosphor persistence

Various phosphors are available depending upon the needs of the measurement or display application. The brightness, color, and persistence of the illumination depends upon the type of phosphor used on the CRT screen. Phosphors are available with persistences ranging from less than one microsecond to several seconds. For visual observation of brief transient events, a long persistence phosphor may be desirable. For events which are fast and repetitive, or high frequency, a short-persistence phosphor is generally preferable.

Microchannel plate

When displaying fast one-shot events the electron beam must deflect very quickly, with few electrons impinging on the screen; leading to a faint or invisible image on the display. Oscilloscope CRTs designed for very fast signals can give a brighter display by passing the electron beam through a micro-channel plate just before it reaches the screen. Through the phenomenon of secondary emission this plate multiplies the number of electrons reaching the phosphor screen, giving a significant improvement in writing rate (brightness), and improved sensitivity and spot size as well.

Graticules

Most oscilloscopes have a graticule as part of the visual display, to facilitate measurements. The graticule may be permanently marked inside the face of the CRT, or it may be a transparent external plate made of glass or acrylic plastic. An internal graticule eliminates parallax error, but cannot be changed to accommodate different types of measurements. Oscilloscopes commonly provide a means for the graticule to be illuminated from the side, which improves its visibility.

Image Storage Tubes

These are found in analog phosphor storage oscilloscopes. These are distinct from digital storage oscilloscopes which rely on solid state digital memory to store the image. Where a single brief event is monitored by an oscilloscope, such an event will be displayed by a conventional tube only while it actually occurs. The use of a long persistence phosphor may allow the image to be observed after the event, but only for a few seconds at best. This limitation can be overcome by the use of a direct view storage cathode ray tube (storage tube).

A storage tube will continue to display the event after it has occurred until such time as it is erased. A storage tube is similar to a conventional tube except that it is equipped with a metal grid coated with a dielectric layer located immediately behind the phosphor screen. An externally applied voltage to the mesh initially ensures that the whole mesh is at a constant potential. This mesh is constantly exposed to a low velocity electron beam from a 'flood gun' which operates independently of the main gun. This flood gun is not deflected like the main gun but constantly 'illuminates' the whole of the storage mesh. The initial charge on the storage mesh is such as to repel the electrons from the flood gun which are prevented from striking the phosphor screen.

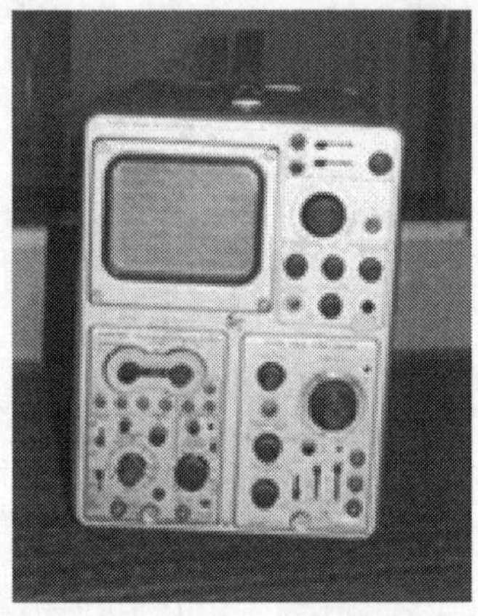

Fig. The Model 564 First Mass Produced Analog Phosphor Storage Oscilloscope

When the main electron gun writes an image to the screen, the energy in the main beam is sufficient to create a 'potential relief' on the storage mesh. The areas where this relief is created no longer repel the electrons from the flood gun which now pass through the mesh and illuminate the phosphor screen. Consequently, the image that was briefly traced out by the main gun continues to be displayed after it has occurred.

The image can be 'erased' by resupplying the external voltage to the mesh restoring its constant potential. The time for which the image can be displayed was limited because, in practice, the flood gun slowly neutralises the charge on the storage mesh. One way of allowing the image to be retained for longer is temporarily to turn off the flood gun. It is then possible for the image to be retained for several days.

The majority of storage tubes allow for a lower voltage to be applied to the storage mesh which slowly restores the initial charge state. By varying this voltage a variable persistence is obtained. Turning off the flood gun and the voltage supply to the storage mesh allows such a tube to operate as a conventional oscilloscope tube.

DATA STORAGE TUBES

Color CRTs

Color tubes use three different phosphors which emit red, green, and blue light respectively. They are packed together in stripes (as inaperture grille designs) or clusters called "triads" (as in shadow mask CRTs). Color CRTs have three electron guns, one for each primary color, arranged either in a straight line or in an equilateral triangular configuration (the guns are usually constructed as a single unit).

Fig. Magnified view of a delta-gun shadow mask color CRT

Fig. Magnified view of aTrinitron color CRT

The triangular configuration is often called "delta-gun", based on its relation to the shape of the Greek letter delta. A grille or mask absorbs the electrons that would otherwise hit the wrong phosphor. A shadow mask tube uses a metal plate with tiny holes, placed so that the electron beam only illuminates the correct phosphors on the face of the tube; the holes are tapered so that the electrons that strike the inside of any hole will be reflected back, if they are not absorbed (e.g. due to local charge accumulation), instead of bouncing through the hole to strike a random (wrong) spot on the screen. Another type of color CRT uses an aperture grille of tensioned vertical wires to achieve the same result.

Convergence and Purity in Color CRTs

Due to limitations in the dimensional precision with which CRTs can be manufactured economically, it has not been practically possible to build color CRTs in which three electron beams could be aligned to hit phosphors of respective color in acceptable coordination, solely on the basis of the geometric configuration of the electron gun axes and gun aperture positions, shadow mask apertures, etc. The shadow mask ensures that one beam will only hit spots certain colors of phosphors, but minute variations in physical alignment of the internal parts among individual CRTs will cause variations in the exact alignment of the beams through the shadow mask, allowing some electrons from, for example, the red beam to hit, say, blue phosphors, unless some individual compensation is made for the variance among individual tubes.

Color convergence and color purity are two aspects of this single problem. Firstly, for correct color rendering it is necessary that regardless of where the beams are deflected on the screen, they hit the same spot (and nominally pass through the same hole or slot) on the shadow mask. This is called convergence. More specifically, the convergence at the center of the screen (with no deflection field applied by the yoke) is called static convergence, and the convergence over the rest of the screen area is called dynamic convergence. The beams may converge at the center of the screen and yet stray from each other as they are deflected toward the edges; such a CRT would be said to have good static convergence but poor dynamic convergence. Secondly, each beam must only strike the phosphors of the color it is intended to strike and no others.

This is called purity. Like convergence, there is static purity and dynamic purity, with the same meanings of "static" and "dynamic" as for convergence. Convergence and purity are distinct parameters; a CRT could have good purity but poor convergence, or vice versa. Poor convergence causes color "shadows" or "ghosts" along displayed edges and contours, as if the image on the screen were intaglio printed with poor registration. Poor purity causes objects on the screen to appear off-color while their edges remain sharp. Purity and convergence problems can occur at the same time, in the same or different areas of the screen or both over the whole screen, and either uniformly or to greater or lesser degrees over different parts of the screen.

The solution to the static convergence and purity problems is a set of color alignment magnets installed around the neck of the CRT. These movable weak permanent magnets are usually mounted on the back end of the deflection yoke assembly and are set at the factory to compensate for any static purity and convergence errors that are intrinsic to the unadjusted tube. Typically there are two or three pairs of two magnets in the form of rings made of plastic impregnated with a magnetic material, with theirmagnetic fields parallel to the planes of the magnets, which are perpendicular to the electron gun axes. Each pair of magnetic rings forms a single effective magnet whose field vector can be fully and freely adjusted (in both direction and magnitude).

By rotating a pair of magnets relative to each other, their relative field alignment can be varied, adjusting the effective field strength of the pair. (As they rotate relative to each other, each magnet's field can be considered to have two opposing components at right angles, and these four components [two each for two magnets] form two pairs, one pair reinforcing each other and the other pair opposing and canceling each other. Rotating away from alignment, the magnets' mutually reinforcing field components decrease as they are traded for increasing opposed, mutually cancelling components.) By rotating a pair of magnets together, preserving the relative angle between them, the direction of their collective magnetic field can be varied. Overall, adjusting all of the convergence/purity magnets allows a finely tuned slight electron beam deflection and/or lateral offset to be applied, which compensates for minor static convergence and purity errors intrinsic to the uncalibrated tube. Once set, these magnets are usually glued in place, but normally they can be freed and readjusted in the field (e.g. by a TV repair shop) if necessary.

On some CRTs, additional fixed adjustable magnets are added for dynamic convergence and/or dynamic purity at specific points on the screen, typically near the corners or edges. Further adjustment of dynamic convergence and purity typically cannot be done passively, but requires active compensation circuits. Dynamic color convergence and purity are one of the main reasons why until late in their history, CRTs were long-necked (deep) and had biaxially curved faces; these geometric design characteristics are necessary for intrinsic passive dynamic color convergence and purity. Only starting around the 1990s did sophisticated active dynamic convergence compensation circuits become available that made short-necked and flat-faced CRTs workable.

These active compensation circuits use the deflection yoke to finely adjust beam deflection according to the beam target location. The same techniques (and major circuit components) also make possible the adjustment of display image rotation, skew, and other complex raster geometry parameters through electronics under user control.

Degaussing

If the shadow mask becomes magnetized, its magnetic field deflects the electron beams passing through it, causing color purity distortion as the beams bend through the mask holes and hit some phosphors of a color other than that which they are intended to strike; e.g. some electrons from the red beam may hit blue phosphors, giving pure red parts of the image a magenta tint. (Magenta is the additive combination of red and blue.) This effect is localized to a specific area of the screen if the magnetization of the shadow mask is localized. Therefore, it is important that the shadow mask is unmagnetized. (A magnetized aperture grille has a similar effect, and everything stated in this subsection about shadow masks applies as well to aperture grilles.)

Most color CRT displays, i.e. television sets and computer monitors, each have a built-in degaussing (demagnetizing) circuit, the primary component of which is a degaussing coil which is mounted around the perimeter of the CRT face inside the bezel. Upon power-up of the CRT display, the degaussing circuit produces a brief, alternating current through the degaussing coil which smoothly decays in strength (fades out) to zero over a period of a few seconds, producing a decaying alternating magnetic field from the coil. This degaussing field is strong enough to remove shadow mask magnetization in most cases. In unusual cases of strong magnetization where the internal degaussing field is not sufficient, the shadow mask may be degaussed externally with a stronger portable degausser or demagnetizer. However, an excessively strong magnetic field, whether alternating or constant, may mechanically deform (bend) the shadow mask, causing a permanent color distortion on the display which looks very similar to a magnetization effect.

The degaussing circuit is often built of a thermo-electric (not electronic) device containing a small ceramic heating element and a positive thermal coefficient (PTC) resistor, connected directly to the switched AC power line with the resistor in series with the degaussing coil. When the power is switched on, the heating element heats the PTC resistor, increasing its resistance to a point where degaussing current is minimal, but not actually zero. In older CRT displays, this low-level current (which produces no significant degaussing field) is sustained along with the action of the heating element as long as the display remains switched on. To repeat a degaussing cycle, the CRT display must be switched off and left off for at least several seconds to reset the degaussing circuit by allowing the PTC resistor to cool to the ambient temperature; switching the display off and immediately back on will result in a weak degaussing cycle or effectively no degaussing cycle.

This simple design is effective and cheap to build, but it wastes some power continuously. Later models, especially Energy Star rated ones, use a relay to switch the entire degaussing circuit on and off, so that the degaussing circuit uses energy only when it is functionally active and needed. The relay design also enables degaussing on user demand through the unit's front panel controls, without switching the unit off and on again. This relay can often be heard clicking off at the end of the degaussing cycle a few seconds after the monitor is turned on, and on and off during a manually-initiated degaussing cycle.

VECTOR MONITORS

Vector monitors were used in early computer aided design systems and in some late-1970s to mid-1980s arcade games such as Asteroids. They draw graphics point-to-point, rather than scanning a raster. Either monochrome or color CRTs can be used in vector displays, and the essential principles of CRT design and operation are the same for either type of display; the main difference is in the beam deflection patterns and circuits.

CRT RESOLUTION

Dot pitch defines the maximum resolution of the display, assuming delta-gun CRTs. In these, as the scanned resolution approaches the dot pitch resolution, moiré appears, as the detail being displayed is finer than what the shadow mask can render. Aperture grille monitors do not suffer from vertical moiré, however, because their phosphor stripes have no vertical detail. In smaller CRTs, these strips maintain position by themselves, but larger aperture grille CRTs require one or two crosswise (horizontal) support strips.

GAMMA

CRTs have a pronounced triode characteristic, which results in significant gamma (a nonlinear relationship in an electron gun between applied video voltage and beam intensity).

OTHER TYPES OF CRTS

Cat's Eye

In better quality tube radio sets a tuning guide consisting of a phosphor tube was used to aid the tuning adjustment. This was also known as a "Magic Eye" or "Tuning Eye". Tuning would be adjusted until the width of a radial shadow was minimized. This was used instead of a more expensive electromechanical meter, which later came to be used on higher-end tuners when transistor sets lacked the high voltage required to drive the device. The same type of device was used with tape recorders as a recording level meter.

Charactrons

Some displays for early computers (those that needed to display more text than was practical using vectors, or that required high speed for photographic output) usedCharactron CRTs. These incorporate a perforated metal character mask (stencil), which shapes a wide electron beam to form a character on the screen. The system selects a character on the mask using one set of deflection circuits, but that causes the extruded beam to be aimed off-axis, so a second set of deflection plates has to re-aim the beam so it is headed toward the center of the screen. A third set of plates places the character wherever required. The beam is unblanked (turned on) briefly to draw the character at that position. Graphics could be drawn by selecting the position on the mask corresponding to the code for a space (in practice, they were simply not drawn), which had a small round hole in the center; this effectively disabled the character mask, and the system reverted to regular vector behavior. Charactrons had exceptionally long necks, because of the need for three deflection systems.

Nimo

Nimo was the trademark of a family of small specialised CRTs manufactured by Industrial Electronics Engineers. These had 10 electron guns which produced electron beams in the form of digits in a manner similar to that of the charactron.

Fig. Nimo tube BA0000-P31

The tubes were either simple single-digit displays or more complex 4- or 6- digit displays produced by means of a suitable magnetic deflection system. Having little of the complexities of a standard CRT, the tube required a relatively simple driving circuit, and as the image was projected on the glass face, it provided a much wider viewing angle than competitive types (e.g., nixie tubes).

Williams Tube

The Williams tube or Williams-Kilburn tube was a cathode ray tube used to electronically store binary data. It was used in computers of the 1940s as a random-access digital storage device. In contrast to other CRTs in this article, the Williams tube was not a display device, and in fact could not be viewed since a metal plate covered its screen.

Zeus thin CRT Display

In the late 1990s and early 2000s Philips Research Laboratories experimented with a type of thin CRT known as the Zeus display which contained CRT-like functionality in aflat panel display. The devices were demonstrated but never marketed.

THE FUTURE OF CRT TECHNOLOGY

Demise

Although a mainstay of display technology for decades, CRT-based computer monitors and televisions constitute a dead technology. The demand for CRT screens has dropped precipitously since 2000, and this falloff had accelerated in the latter half of that decade. The rapid advances and falling prices of LCD flat panel technology, first for computer monitors and then for televisions, has been the key factor in the demise of competing display technologies such as CRT, rear-projection, and plasma display. The end of most high-end CRT production by around 2010 (including high-end Sony and Mitsubishi product lines) means an erosion of the CRT's capability. In Canada and the United States, the sale and production of high-end CRT TVs (30-inch screens) in these markets had all but ended by 2007; just a couple of years later, inexpensive combo CRT TVs (20-inch screens with an integrated VHS or DVD player) have disappeared from discount stores.

It has been common to replace CRT-based televisions and monitors in as little as 5–6 years, although they generally are capable of satisfactory performance for a much longer time. Companies are responding to this trend. Electronics retailers such as Best Buy have been steadily reducing store spaces for CRTs.

In 2005, Sony announced that they would stop the production of CRT computer displays. Samsung did not introduce any CRT models for the 2008 model year at the 2008 Consumer Electronics Show and on 4 February 2008 Samsung removed their 30" wide screen CRTs from their North American website and has not replaced them with new models. The demise of CRT, however, has been happening more slowly in the developing world. According to iSupply, production in units of CRTs was not surpassed by LCDs production until 4Q 2007, owing largely to CRT production at factories in China.

In the United Kingdom, DSG (Dixons), the largest retailer of domestic electronic equipment, reported that CRT models made up 80–90% of the volume of televisions sold at Christmas 2004 and 15–20% a year later, and that they were expected to be less than 5% at the end of 2006. Dixons ceased selling CRT televisions in 2007.

Causes

CRTs, despite recent advances, have remained relatively heavy and bulky and take up a lot of space in comparison to other display technologies. CRT screens have much deeper cabinets compared to flat panels and rear-projection displays for a given screen size, and so it becomes impractical to have CRTs larger than 40 inches (102 cm).

The CRT disadvantages became especially significant in light of rapid technological advancements in LCD and plasma flat-panels which allow them to easily surpass 40 inches (102 cm) as well as being thin and wall-mountable, two key features that were increasingly being demanded by consumers.

By 2006, although the price points of CRTs were generally much lower than LCD and plasma flat panels, large screen CRTs (30-inches or more) were as expensive as a similar-sized LCD.

Monochrome CRTs use less power than color CRTs (but they are not more efficient overall). This is because up to 2/3 of the backlight power of LCD and rear-projection CRT displays are lost to the RGB stripe filter. Older LCDs also have poorer color rendition and can change color with viewing angle, though modern PVA and IPSLCDs have greatly attenuated these problems.

Slimmer CRT

Some CRT manufacturers, both LG Display and Samsung Display, have innovated CRT technology by creating a slimmer tube. Slimmer CRT has a trade name Superslim and Ultraslim. A 21 inch flat CRT has 447.2 millimeter depth. The depth of Superslim is 352 millimeters and Ultraslim is 295.7 millimeters.

Fig. A comparison between 21 inch Superslim and Ultraslim CRT

Resurgence in Specialized Markets

In the first quarter of 2008, CRTs retook the #2 technology position in North America from plasma, due to the decline and consolidation of plasma display manufacturers. DisplaySearch has reported that although in the 4Q of 2007 LCDs surpassed CRTs in worldwide sales, CRTs then outsold LCDs in the 1Q of 2008. CRTs are useful for displaying photos with high pixels per unit area and correct color balance. LCDs, as currently the most common flatscreen technology, have generally inferior color rendition (despite having greater overall brightness) due to thefluorescent lights commonly used as a backlight. CRTs are still popular in the printing and broadcasting industries as well as in the professional video, photography, and graphics fields due to their greater color fidelity, contrast, and better viewing from off-axis (wider viewing angle). CRTs also still find adherents in video gaming because of their higher resolution per initial cost, lowest possible input lag, fast response time, and multiple native resolutions. CRT monitors are still widely used in the study of the brain's visual processing (e.g. in psychophysics). The speed and fidelity of their response, combined with the simplicity of their design, makes them well-suited for experiments where scientists need to have very fine control over stimuli which are presented to an observer.

HEALTH CONCERNS

Ionizing Radiation

CRTs can emit a small amount of X-ray radiation as a result of the electron beam's bombardment of the shadow mask/aperture grille and phosphors. The amount of radiation escaping the front of the monitor is widely considered unharmful. The Food and Drug Administration regulations in 21 C.F.R. 1020.10 are used to strictly limit, for instance, television receivers to 0.5 milliroentgens per hour (mR/h) (0.13 μC/(kg.h) or 36 pA/kg) at a distance of 5 cm (2 in) from any external surface; since 2007, most CRTs have emissions that fall well below this limit.

Toxicity

Older color and monochrome CRTs may contain toxic substances, such as cadmium, in the phosphors. The rear glass tube of modern CRTs may be made fromleaded glass, which represent an environmental hazard if disposed of improperly. By the time personal computers were produced, glass in the front panel (the viewable portion of the CRT) used barium rather than lead, though the rear of the CRT was still produced from leaded glass. Monochrome CRTs typically do not contain enough leaded glass to fail EPA TCLP tests. While the TCLP process grinds the glass into fine particles in order to expose them to weak acids to test for leachate, intact CRT glass does not leache (The lead is vitrified, contained inside the glass itself, similar to leaded glass crystalware). In October 2001, the United States Environmental Protection Agency created rules stating that CRTs must be brought to special recycling facilities. In November 2002, the EPA began fining companies that disposed of CRTs through landfills or incineration. Regulatory agencies, local and statewide, monitor the disposal of CRTs and other computer equipment. In Europe, disposal of CRT televisions and monitors is covered by the WEEE Directive.

Flicker

At low refresh rates (60 Hz and below), the periodic scanning of the display may produce an irritating flicker that some people perceive more easily than others, especially when viewed with peripheral vision. Flicker is commonly associated with CRT as most televisions run at 50 Hz (PAL) or 60 Hz (NTSC), although there are some 100 Hz PAL televisions that are flicker-free. Typically only low-end monitors run at such low frequencies, with most computer monitors supporting at least 75 Hz and high-end monitors capable of 100 Hz or more to eliminate any perception of flicker. Non-computer CRTs or CRT for sonar or radar may have long persistence phosphor and are thus flicker free. If the persistence is too long on a video display, moving images will be blurred.

High-Frequency Audible Noise

50 Hz/60 Hz CRTs used for television operate with horizontal scanning frequencies of 15,734 Hz (for NTSC systems) or 15,625 Hz (for PAL systems).

These frequencies are at the upper range of human hearing and are inaudible to many people; however, some people (especially children) will perceive a high-pitched tone near an operating television CRT. The sound is due to magnetostriction in the magnetic core and periodic movement of windings of the flyback transformer. This problem does not occur on 100/120 Hz TVs and on non-CGA computer displays, because they are working on much higher frequencies (22 kHz to >100 kHz) compared to the low-frequency noise (50 Hz or 60 Hz) of mains hum.

Implosion

High vacuum inside glass-walled cathode ray tubes permits electron beams to fly freely—without colliding into molecules of air or other gas. If the glass is damaged, atmospheric pressure can collapse the vacuum tube into dangerous fragments which accelerate inward and then spray at high speed in all directions. The implosion energy is proportional to the evacuated volume of the CRT. Although modern cathode ray tubes used in televisions and computer displays have epoxy-bonded face-plates or other measures to prevent shattering of the envelope, CRTs must be handled carefully to avoid personal injury.

SECURITY CONCERNS

Under some circumstances, the signal radiated from the electron guns, scanning circuitry, and associated wiring of a CRT can be captured remotely and used to reconstruct what is shown on the CRT using a process called Van Eck phreaking. Special TEMPEST shielding can mitigate this effect. Such radiation of a potentially exploitable signal, however, occurs also with other display technologies and with electronics in general.

Recycling

As electronic waste, CRTs are considered one of the hardest types to recycle. CRTs have relatively high concentration of lead and phosphors (not phosphorus), both of which are necessary for the display. There are several companies in the United States that charge a small fee to collect CRTs, then subsidize their labour by selling the harvested copper, wire, and printed circuit boards. The United States Environmental Protection Agency (EPA) includes discarded CRT monitors in its category of "hazardous household waste" but considers CRTs that have been set aside for testing to be commodities if they are not discarded, speculatively accumulated, or left unprotected from weather and other damage. Leaded CRT glass is sold to be remelted into other CRTs, or even broken down and used in road construction.

COMMUNICATIONS SATELLITE

A communications satellite or comsat is an artificial satellite sent to space for the purpose of telecommunications. Modern communications satellites use a variety of orbits including geostationary orbits, Molniya orbits, elliptical orbits and low (polar and non-polar Earth orbits). For fixed (point-to-point) services, communications satellites provide a microwave radio relay technology complementary to that of communication cables. They are also used for mobile applications such as communications to ships, vehicles, planes and hand-held terminals, and for TV and radio broadcasting.

HISTORY

The Merriam-Webster dictionary defines a satellite as a celestial body orbiting another of larger size or a manufactured object or vehicle intended to orbit the earth, the moon, or another celestial body. Today's satellite communications can trace their origins all the way back to the Moon. A project named Communication Moon Relay was a telecommunication project carried out by the United States Navy. Its objective was to develop a secure and reliable method of wireless communication by using the Moon as a natural communications satellite.

The first artificial satellite used solely to further advances in global communications was a balloon named Echo 1. Echo 1 was the world's first artificial communications satellite capable of relaying signals to other points on Earth. It soared 1,000 miles (1,609 km) above the planet after its Aug. 12, 1960 launch, yet relied on humanity's oldest flight technology — ballooning. Launched by NASA, Echo 1 was a giant metallic balloon 100 feet (30 meters) across. The world's first inflatable satellite — or "satelloon," as they were informally known — helped lay the foundation of today's satellite communications. The idea behind a communications satellite is simple: Send data up into space and beam it back down to another spot on the globe. Echo 1 accomplished this by essentially serving as an enormous mirror 10 stories tall that could be used to bounce communications signals off of.

The first American satellite to relay communications was Project SCORE in 1958, which used a tape recorder to store and forward voice messages. It was used to send a Christmas greeting to the world from U.S. President Dwight D. Eisenhower. NASA launched the Echo satellite in 1960; the 100-foot (30 m) aluminised PET film balloon served as a passive reflector for radio communications. Courier 1B, built by Philco, also launched in 1960, was the world's first active repeater satellite.

It is commonly believed that the first "satellite" was Sputnik 1. Put into orbit by the Soviet Union on October 4, 1957, it was equipped with an onboard radio-transmitter that worked on two frequencies: 20.005 and 40.002 MHz. Sputnik 1 was launched as a step in the exploration of space and rocket development. While incredibly important it was not placed in orbit for the purpose of sending data from one point on earth to another. Hence, it was not the first "communications" satellite, but it was the first artificial satellite in the steps leading to today's satellite communications. Telstar was the first active, direct relay communications satellite. Belonging to AT&T as part of a multi-national agreement between AT&T, Bell Telephone Laboratories, NASA, the British General Post Office, and the French National PTT (Post Office) to develop satellite communications, it was launched by NASA from Cape Canaveral on July 10, 1962, the first privately sponsored space launch. Relay 1 was launched on December 13, 1962, and became the first satellite to broadcast across the Pacific on November 22, 1963. An immediate antecedent of the geostationary satellites was Hughes' Syncom 2, launched on July 26, 1963. Syncom 2 revolved around the earth once per day at constant speed, but because it still had north-south motion, special equipment was needed to track it.

Geostationary orbits

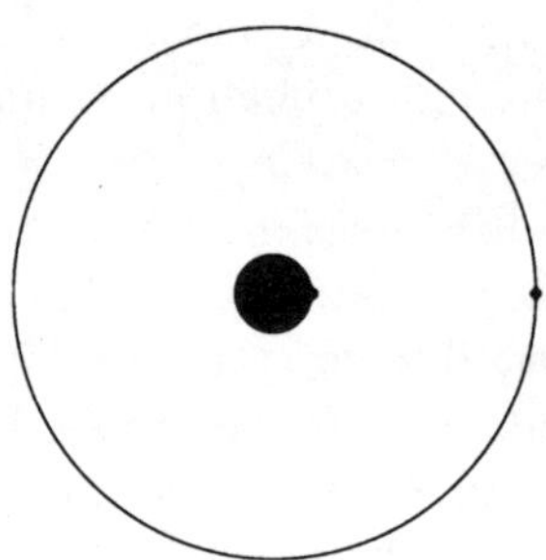

Fig. Geostationary orbit

To an observer on the earth, a satellite in a geostationary orbit appears motionless, in a fixed position in the sky. This is because it revolves around the earth at the earth's own angular velocity (360 degrees every 24 hours, in an equatorial orbit). A geostationary orbit is useful for communications because ground antennas can be aimed at the satellite without their having to track the satellite's motion. This is relatively inexpensive. In applications that require a large number of ground antennas, such asDirectTV distribution, the savings in ground equipment can more than outweigh the cost and complexity of placing a satellite into orbit.

The concept of the geostationary communications satellite was first proposed by Arthur C. Clarke, building on work by Konstantin Tsiolkovsky and on the 1929 work by Herman Potocnik (writing as Herman Noordung) Das Problem der Befahrung des Weltraums — der Raketen-motor. In October 1945 Clarke published an article titled "Extra-terrestrial Relays" in the British magazine Wireless World.

The article described the fundamentals behind the deployment of artificial satellites in geostationary orbits for the purpose of relaying radio signals. Thus, Arthur C. Clarke is often quoted as being the inventor of the communications satellite.

The first geostationary satellite was Syncom 3, launched on August 19, 1964, and used for communication across the Pacific starting with television coverage of the 1964 Summer Olympics. Shortly after Syncom 3, Intelsat I, aka Early Bird, was launched on April 6, 1965 and placed in orbit at 28° west longitude. It was the first geostationary satellite for telecommunications over the Atlantic Ocean. On November 9, 1972, Canada's first geostationary satellite serving the continent, Anik A1, was launched by Telesat Canada, with the United States following suit with the launch of Westar 1 by Western Union on April 13, 1974. On May 30, 1974, the first geostationary communications satellite in the world to be three-axis stabilized was launched: the experimental satellite ATS-6 built for NASA

After the launches of the Telstar through Westar 1 satellites, RCA Americom (later GE Americom, now SES) launched Satcom 1 in 1975. It was Satcom 1 that was instrumental in helping early cable TV channels such as WTBS (now TBS Superstation), HBO, CBN (now ABC Family), and The Weather Channel become successful, because these channels distributed their programming to all of the local cable TV headends using the satellite. Additionally, it was the first satellite used by broadcast television networks in the United States, like ABC, NBC, and CBS, to distribute programming to their local affiliate stations. Satcom 1 was widely used because it had twice the communications capacity of the competing Westar 1 in America (24 transponders as opposed to the 12 of Westar 1), resulting in lower transponder-usage costs. Satellites in later decades tended to have even higher transponder numbers. By 2000, Hughes Space and Communications (now Boeing Satellite Development Center) had built nearly 40 percent of the more than one hundred satellites in service worldwide. Other major satellite manufacturers include Space Systems/Loral, Orbital Sciences Corporation with the STAR Bus series, Indian Space Research Organization,Lockheed Martin (owns the former RCA Astro Electronics/GE Astro Space business), Northrop Grumman, Alcatel Space, now Thales Alenia Space, with the Spacebus series, and Astrium.

Low-Earth-orbiting satellites

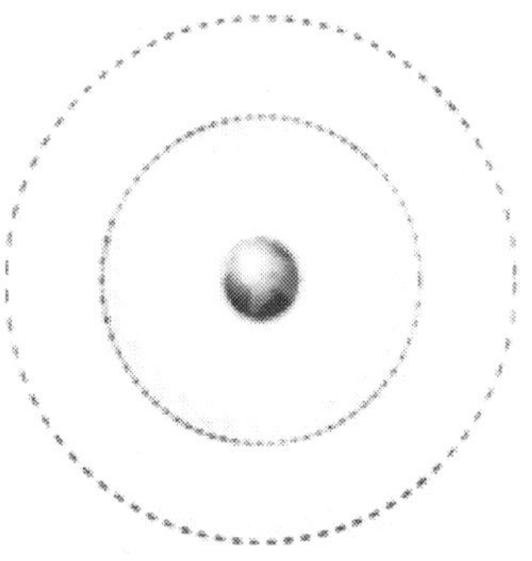

Fig. Low Earth orbit in Cyan

A low Earth orbit (LEO) typically is a circular orbit about 200 kilometres (120 mi) above the earth's surface and, correspondingly, a period (time to revolve around the earth) of about 90 minutes. Because of their low altitude, these satellites are only visible from within a radius of roughly 1000 kilometers from the sub-satellite point. In addition, satellites in low earth orbit change their position relative to the ground position quickly. So even for local applications, a large number of satellites are needed if the mission requires uninterrupted connectivity.

Low-Earth-orbiting satellites are less expensive to launch into orbit than geostationary satellites and, due to proximity to the ground, do not require as high signal strength (Recall that signal strength falls off as the square of the distance from the source, so the effect is dramatic). Thus there is a trade off between the number of satellites and their cost. In addition, there are important differences in the onboard and ground equipment needed to support the two types of missions.

A group of satellites working in concert is known as a satellite constellation. Two such constellations, intended to provide satellite phone services, primarily to remote areas, are the Iridium and Globalstar systems. The Iridium system has 66 satellites. It is also possible to offer discontinuous coverage using a low-Earth-orbit satellite capable of storing data received while passing over one part of Earth and transmitting it later while passing over another part. This will be the case with the CASCADE system ofCanada's CASSIOPE communications satellite. Another system using this store and forward method is Orbcomm.

Molniya Satellites

Geostationary satellites must operate above the equator and therefore appear lower on the horizon as the receiver gets the farther from the equator. This will cause problems for extreme northerly latitudes, affecting connectivity and causing multipath (interference caused by signals reflecting off the ground and into the ground antenna). For areas close to the North (and South) Pole, a geostationary satellite may appear below the horizon. Therefore Molniya orbit satellite have been launched, mainly in Russia, to alleviate this problem. The first satellite of the Molniya series was launched on April 23, 1965 and was used for experimental transmission of TV signal from a Moscow uplinkstation to downlink stations located in Siberia and the Russian Far East, in Norilsk, Khabarovsk, Magadan and Vladivostok. In November 1967 Soviet engineers created a unique system of national TV network of satellite television, called Orbita, that was based on Molniya satellites.

Molniya orbits can be an appealing alternative in such cases. The Molniya orbit is highly inclined, guaranteeing good elevation over selected positions during the northern portion of the orbit. (Elevation is the extent of the satellite's position above the horizon. Thus, a satellite at the horizon has zero elevation and a satellite directly overhead has elevation of 90 degrees). The Molniya orbit is designed so that the satellite spends the great majority of its time over the far northern latitudes, during which its ground footprint moves only slightly. Its period is one half day, so that the satellite is available for operation over the targeted region for six to nine hours every second revolution. In this way a constellation of three Molniya satellites (plus in-orbit spares) can provide uninterrupted coverage.

DECODER

Fig. A Digitrax DH163AT DCC decoder in anAthearn locomotive before the shell goes on.

A decoder is a device which does the reverse operation of an encoder, undoing the encoding so that the original information can be retrieved. The same method used to encode is usually just reversed in order to decode. It is a combinational circuit that converts binary information from n input lines to a maximum of 2n unique output lines.

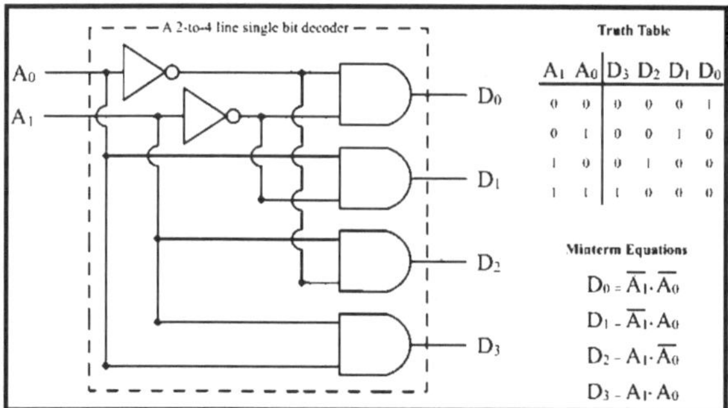

A_1	A_0	D_3	D_2	D_1	D_0
0	0	0	0	0	1
0	1	0	0	1	0
1	0	0	1	0	0
1	1	1	0	0	0

In digital electronics, a decoder can take the form of a multiple-input, multiple-outputlogic circuit that converts coded inputs into coded outputs, where the input and output codes are different. e.g. n-to-2n, binary-coded decimal decoders. Enable inputs must be on for the decoder to function, otherwise its outputs assume a single "disabled" output code word. Decoding is necessary in applications such as data multiplexing, 7 segment display and memory address decoding. The example decoder circuit would be an AND gate because the output of an AND gate is "High" (1) only when all its inputs are "High." Such output is called as "active High output". If instead of AND gate, the NAND gate is connected the output will be "Low" (0) only when all its inputs are "High". Such output is called as "active low output". A slightly more complex decoder would be the n-to-2n type binary decoders. These type of decoders are combinational circuits that convert binary information from 'n' coded inputs to a maximum of 2n unique outputs. We say amaximum of 2n outputs because in case the 'n' bit coded information has unused bit combinations, the decoder may have less than 2n outputs. We can have 2-to-4 decoder, 3-to-8 decoder or 4-to-16 decoder. We can form a 3-to-8 decoder from two 2-to-4 decoders (with enable signals).

Similarly, we can also form a 4-to-16 decoder by combining two 3-to-8 decoders. In this type of circuit design, the enable inputs of both 3-to-8 decoders originate from a 4th input, which acts as a selector between the two 3-to-8 decoders. This allows the 4th input to enable either the top or bottom decoder, which produces outputs of D(0) through D(7) for the first decoder, and D(8) through D(15) for the second decoder. A decoder that contains enable inputs is also known as a decoder-demultiplexer. Thus, we have a 4-to-16 decoder produced by adding a 4th input shared among both decoders, producing 16 outputs.

Row Select

Most kinds of random-access memory use a n-to-2n decoder to convert the selected address on the address bus to one of the row address select lines.

Instruction Decoder

In CPU design, the instruction decoder is the part of the CPU that converts the bits stored in the instruction register – or, in CPUs that have microcode, the microinstruction – into the control signals that control the other parts of the CPU. A simple CPU with 8 registers may use 3-to-8 logic decoders inside the instruction decoder to select two source registers of the register file to feed into the ALU as well as the destination register to accept the output of the ALU. A typical CPU instruction decoder also includes several other things.

DIGIBOX

Fig. An Amstrad Digibox with Viewing Card Inserted

The Digibox is a device marketed by British Sky Broadcasting in the UK and Ireland to enable home users to receive digital satellite television broadcasts (satellite receiver) from the Astrasatellites at 28.2° east. An internet service is also available through the device, similar in some ways to the American MSN TV. The first Digiboxes shipped to consumers in mid-1998, and the hardware reference design is unchanged since. Compared to other satellite receivers, they are severely restricted.

Base technical details

The Digibox's internal hardware details are not publicly disclosed, however some details are clearly visible on the system. All early boxes except the Pace Javelin feature dualSCART outputs, an RS232 serial port, a dual-output RF modulator with passthrough, and RCA socketed audio outputs, as well as a 33.6 modem and an LNB cable socket. AVideoGuard card slot, as well as a second smart-card reader are fitted to the front (these are for the Sky viewing card and other interactive cards). All share an identical user interface and EPG, with the exception of Sky+ HD boxes which use the new Sky+ HD Guide.

All Sky and early HD boxes had a s-video output socket. The latest DRX595 has dropped the RF modulator outputs. A PC type interface was fitted internally to some early standard boxes but was never utilised by Sky. The latest HD boxes only have a single scart socket but have a RCA/phono socket for composite video output.

All Sky+ and HD boxes have an optical sound output. The serial port outputs data used for the Sky Gnome and Sky Talker. The Sky Gamepad sends data to the box via the serial port.

Uniquely, the second RF port outputs a 9 V power signal which is used to power 'tvLINK' devices that can be attached to the RF cable next to a TV in a remote room. The tvLINK has an IR detector and sends commands from the remote control back to the digibox on a 6 MHz carrier on the RF cable. This allows the sky box to be viewed and controlled from another room by running a single RF cable.

All Digiboxes used to run on OpenTV (the latest HD boxes now use what is known internally as Project Darwin software) with Sky's EPG software and NDS VideoGuardconditional access.

The Digibox receives software updates over the air, even when in standby mode should an update be available. The software features Sky-controlled channel numbering, the ability to view OpenTV or WapTV applications provided as "interactive" or "teletext" content on channels, parental controls, the ability to order PPVevents and some basic control over lists of favourite channels and show reminders.

Early decoders seemingly support only 700 channels approximately in their channel listing, as Sky has announced it is halting SD channels' launch applications, with over 100 awaiting launch, over 600 existing channels, and an average closure rate of 1 per week.

Digibox Remote Control

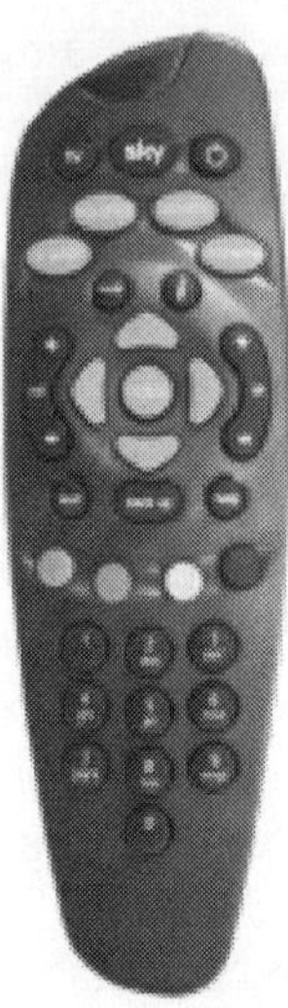

Fig. The Standard Sky Remote in blue

The Digibox remote control comes in four physical designs - blue with new Sky logo, blue with old Sky logo (this version was being issued from around the first year of the Sky service, although that logo fell out of use before it launched), black with new Sky logo (Sony boxes only) and silver with new Sky logo. However, all have multiple variations at present with a new version of Sky remote control being produced every year, to support the addition of new televisions to its universal remote control capabilities.

Use outside Sky's System

The units are DVB-S compatible, and usually carry the DVB logo on the front. However, their use as a DVB-S receiver for anything other than Sky services is seriously limited by their reduced choice of symbol rates (22,000 and 27,500; additionally 28,250 and 29,000 on Sky+ HD), and their inability to store more than 50 (fewer still on some models) non-EPG channels without losing them. Following a software "upgrade" the digibox will not display programmes from non 28.2 East Satellites. Additionally, once any form of parental controls are enabled, the "Other Channels" menu requires PIN entry on every use. The box also refuses to let users view channels which are free-to-air but displaying flags claiming encryption, which locks out some channels even on the satellites Sky use themselves, such as Free to View channels.

Manufacturers and Non-Common Features

Digiboxes have been made by Amstrad, Sony, Thomson, Panasonic, Grundig, and most commonly, Pace. Although the reference designs were identical, a number of digibox lines have specific faults or traits, such as failing modems on Grundig units, and unstable tuners on older model Pace boxes. Some units add features not found on other models, such as S-Video sockets on some Grundig units, and TOSLINK output on Sony's.

For a long period of time in 2004 and 2005, only one Digibox was in production, the Pace DS430N, but Amstrad and Thomson have resumed production with the DRX500 and DSi4212 boxes respectively. Digiboxes for new customers are assigned randomly, and cannot be chosen, creating an after market for specific boxes due to their individual features. Sky bought out Amstrad satellite production in 2007 and now only distribute boxes which initially appeared as Amstrad models but are now badged merely as Sky.

Standardised Design

In late 2005, it was announced that all future Digiboxes would have a standardised cosmetic design, although retain the three current makers, and have a slightly redesigned remote control, which would be recoloured white with some blue keys. Thomson and Amstrad began supplying these boxes in late 2005, with Pace following in late 2006. The standardised design is known as the "Flow" design.

Other Digiboxes

A second generation of Digibox exists, marketed as Sky+. These PVR units have three versions: the 40 GB PVR1, the 40 GB PVR2, and the PVR3 which is fitted with a 160 GB HDD with 80GB available for user recordings, and the remaining 80 GB reserved for use by Sky Anytime. All have dual LNB inputs and an optical digital audio output, as well as all other features of the Sky box. These units are manufactured by Amstrad, Pace and Thomson only, and use a different remote control. USB ports are fitted to the PVR3. When Sky+ was launched, there was an additional £10 monthly charge to access the Sky+ functionality if two or more premium packages were not subscribed to. Sky have now removed this charge, effective 1 July 2007. Without a Sky+ Subscription, the Sky+ box reverts to the functionality of a normal Digibox, albeit without Autoview modes. The Sky+160 box has 160 GB to record to and is manufactured solely by Thomson. This unit is also fitted with USB ports. (this unit has been discontinued) A third generation of Digibox also exists, with the additional ability to receive DVB-S2 HDTV signals in the MPEG-4 format. The initial sole manufacturer of these Sky+ HDboxes was Thomson, and made their debut on 22 May 2006 with the launch of HDTV channels on Sky. Although around 17,000 digiboxes were initially produced, they could not meet demand and some had to wait for longer for their SkyHD Digibox. Past manufacturers are Thomson, Pace, Amstrad and Samsung.

These boxes have all the features of a Sky+ box, as well as a HDMI output, and the early Thomson DSI8215 model has component video output. Both

SATA and Ethernetports are also supplied, currently used for the Sky Anytime+ service. As well as requiring a Sky+ subscription or any subscription, HD versions of the subscription channels require the payment of an additional £10.25 (or €15) per month, although HD channels can be viewed on free-to-air services such as BBC One HD, BBC Two HD, ITV HD,Channel 4 HD, NHK World HD and with a Freesat from Sky card, Channel 5 HD. On 16 March 2011, Sky launched the Sky HD box, primarily targeted at multiroom subscribers. and also used for Freesat from Sky installations. The Sky HD box is not a personal video recorder, meaning it has no hard disk and cannot support the Sky+ functionality.

Card Pairing

All Sky boxes, whether Digiboxes, Sky+ or SkyHD incorporate card pairing. This involves the 'marrying' or 'pairing' of a viewing card to one particular STB, thus preventing one card (and indeed subscription) being used on multiple digiboxes. During installation the engineer will initiate a 'callback' from the STB to Sky via modem and telephone line, transmitting details of the viewing card number and the box in which it is installed. Once this step has been completed it is no longer possible to view premium channels such as Movies or Sports on any other box besides that which the card was paired with. Notably, the card can still be used to view non-premium channels such as Sky1 but will display an error message when attempting to watch a movie channel or sports channel with an unpaired box and card combination. A card can be 're-paired' in some instances such as STB replacement or multiroom relocation, however this must be initiated by Sky and cannot be completed by an end-user.

Power Consumption

Standard Digiboxes use almost as much power in standby as when active; the "standby" setting merely mutes the sound and cuts off the picture, but internal signal processing continues at the same rate. Sky+ boxes are believed to reduce their power consumption more significantly in standby because they can spin down hard disks. Power consumption for the standard box varies from around 10 to 18 Watts. Most Sky+HD boxes consume up to 60W when active, falling to ~30W when the disc is powered down. To some degree, this has been addressed with the latest DRX890/895 HD boxes. These have a power consumption of 45W max but now have a "deep sleep" mode consuming around 1/2 W.

DISH TV

Dish TV (stylised as dishtv), is an Indian direct-broadcast satellite service provider. It is a division of Zee Network Enterprise(Essel Group Venture). It was ranked # 437 and # 5 on the list of media companies in Fortune India 500 roster of India's largest corporations in 2011. It uses MPEG-2 digital compression technology, transmitting using NSS-6 Satellite at 95.0. Dish TV's managing director and Head Of Business is Jawahar Goel who is also the promoter of Essel Group and is also the President of Indian Broadcasting Foundation. Zee Network incorporated dishtv to modernize television (TV) viewing. It provides features such as Electronic Programme Guide (EPG), parental lock, games, 400+ channels and services, interactive TV and movies on demand. Its primary competitors are Tata Sky, Airtel digital TV, Videocon d2h and cable television providers.

History

DTH service was launched in India back in 2004 with the launch of Dish TV by Essel Group's Zee Entertainment Enterprises. Dish TV was the only DTH service provider in India to carry the two Turner channels: Turner Classic Movies and Boomerang. But, both the channels were removed from its platform due to unknown reasons in March 2009. In October 2010 Dish TV added the long awaited Neo Sports and Neo Cricket on its platform

Satellite link

Dish TV uses NSS-6 to broadcast its programmes. NSS-6 was launched on 17 December 2002 by European-based satellite provider, NewSkies. Dish TV hopped on to NSS-6 from an INSAT satellite in July 2004. The change in the satellite was made to increase the channel offering as NSS 6 offered more transponder capacity. However, Dish TV booked additional transponders on the new AsiaSat 5 satellite for starting its MPEG-4 based HD services. Dish TV is currently using 4 transponders on Asiasat 5.

Subscriber Base

As of 31 October 2012, Dish TV had about 13 million customers. Dish TV is presently Asia's largest satellite television provider. Dish TV launched its high definition service called Dish truHD in the year 2010. With this service, subscribers can enjoy 5X picture clarity on their HDTV, a 16:9 wide aspect ratio and 5.1 surround sound.

Dish truHD+

Dish TV Recently introduced its DVR service which requires an External USB Hard disk drive to be plugged into the Set Top Box's USB Port, the DVR can provide & support recording space up to 2 TB.

HD READY

The HD ready is a certification program introduced in 2005 by EICTA (European Information, Communications and Consumer Electronics Technology Industry Associations), now DIGITALEUROPE. There are currently four different labels: "HD ready", "HD TV", "HD ready 1080p", "HD TV 1080p". The logos are awarded to television equipment capable of certain features. In the USA, a similar "HD Ready" term usually refers to any display that is capable of accepting and displaying a high-definition signal at either 720p, 1080i or 1080p using a component video or digital input, but does not have a built-in HD-capable tuner.

History

The "HD ready" certification program was introduced on January 19, 2005. The labels and relevant specifications are based on agreements between over 60 broadcasters and manufacturers of the European HDTV Forum at its second session in June 2004, held at the Betzdorf, Luxembourg headquarters of founding member SES Astra. The "HD ready" logo is awarded to television equipment capable of displaying High Definition (HD) pictures from an external source, however it does not have to feature a digital tuner to decode an HD signal; devices with tuners were certified under a separate "HD TV" logo, which does not require a "HD ready" display device.

Before the introduction of the "HD ready" certification, many TV sources and displays were being promoted as capable of displaying high definition pictures when they were in fact SDTV devices; according to Alexander Oudendijk, senior VP of marketing for Astra, in early 2005 there were 74 different devices being sold as ready for HD that were not. Devices advertised as HD-compatible or HD ready could take HDTV-signal as an input (via analog -YPbPr or digital DVI or HDMI), but they did not have enough pixels for true representation of even the lower HD resolution (1280 × 720) (CRT based sets only capable of SDTV-resolution or VGA-resolution, 640×480 pixels, or the plasma-based sets with 1024 × 768 resolution), much less the higher HD resolution (1920 × 1080), and so were unable to display the HD picture without downscaling to a lower resolution.

Industry-sponsored labels such as "Full HD" were misleading as well, as they can refer to devices which do not fulfil some essential requirements such as having 1:1 pixel-to-pixel mapping with no overscan or accepting a 1080p signal.

A UK BBC television programme found that separate labels for display devices and TV tuners/decoders confused purchasers, many of whom bought HD-ready equipment expecting to be able to receive HD with no additional equipment; they were sometimes actively misled by salespeople—a 2007 Ofcom survey found that 12% were told explicitly that they could view analog SDTV transmissions in HD, 7% that no extra equipment was needed, and 14% that HD-ready sets would receive existing digital SDTV transmissions in HD.

On August 30, 2007, 1080p versions of the logos and licensing agreements were introduced; as an improvement to the earlier scheme, "HD TV 1080p" logo now requires "HD ready 1080p" certification.

Requirements and Logos

HD ready and HD ready 1080p logos are awarded to displays (including integrated television sets, computer monitors and projectors) which have certain capabilities to process and display high-definition source video signal, outlined in a table below. The HD TV logo is awarded to either integrated digital television sets (containing a display conforming to "HD ready" requirements) or standalone set-top boxes which are capable of receiving, decoding and outputting or displaying high-definition broadcasts (that is, include a DVB tuner for cable, terrestrial or satellite broadcasting, a video decoder which supports H.264/MPEG-4 AVC compression in 720p and 1080i signal formats, and either video outputs or an integrated display capable of handling such signals). The HD TV 1080p logo is awarded to integrated digital television sets which have a display conforming to "HD ready 1080p" requirements, a DVB tuner and a decoder capable of processing 1080p signal. In order to be eligible for the "HD ready 1080p" or "HD Ready" logo, a display device has to meet the following requirements:

Requirements	HD ready	HD ready 1080p
Minimum native resolution	720 horizontal lines (rows) in widescreen ratio	1920×1080
Analogue YPbPr HD input	Yes	Yes
Digital HDMI or DVI HD input	Yes	Yes
The HDMI or DVI input supports copy protection (HDCP)	Yes	Yes
720p HD (1280×720 progressive @50 & 60 Hz)	Yes	Yes
1080i HD (1920×1080 interlaced @50 & 60 Hz)	Yes	Yes
1080p HD (1920×1080 progressive @24, 50 & 60 Hz)	No	Yes
Accepted video formats are reproduced without distortion	No	Yes
Display 1080p and 1080i video without overscan (1:1 pixel mapping)	No	Yes
Display native video modes at the same, or higher, refresh rate	No	Yes

Technical References

- DVI: DDWG, "Digital Visual Interface", rev 1.0, April 2, 1999 as further qualified in EIA861B, "A DTV Profile for Uncompressed High Speed Digital Interfaces" May 2002, furthermore allowing both DVI-D and DVI-I connectors, requiring compliance to both 50 and 60 Hz profiles, and requiring support for both 720p and 1080i video formats.
- HDMI: HDMI Licensing, LLC, "High-Definition Multimedia Interface", rev.1.1, May 20, 2004
- HDCP: Intel, "High-Bandwidth Digital Content Protection System", rev 1.1, June 9, 2003. (For DVI input, HDCP rev 1.0 will apply.)
- YPBPR: EIA770.3-A, March 2000, with the notice that the connectors required may be available only through an adaptor.

2

Achievement for JPEG Decoder

INTRODUCTION

Generally speaking, the JPEG decoder is the inverse of JPEG encoder as shown in Fig.

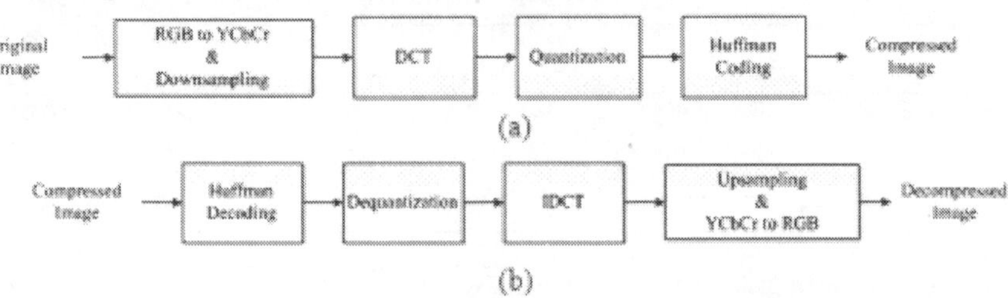

Fig. Flow chart of JPEG encoder and decoder. (a) Encoder, (b) Decoder.

A brief description of the JPEG encoder:

- Reading the pixel value of the original image in RGB color space.
- Transfer the RGB color space to YCbCr space since the human eyes are sensitive to luminance than chrominance which can be downsampled (ex. 4:2:0) to save the storage.
- Transfer the YCbCr space from space domain to frequency domain by discrete cosine transform (DCT) since the human eyes are sensitive to low frequency than high frequency which can be hardly quantized to save the storage.
- Using quantization to decrease the DCT coefficient values required to recode.
- Encode the bitstream by Huffman coding due to the probability distribution of symbols. Symbols with high probability will have shorter codelength, which is a good property to decrease the memory usage.

A brief description of the JPEG decoder:

- Reading the bitstream and decode by Huffman decoding.
- Dequantize the quantized DCT coefficient values.

- Do the inverse DCT (IDCT) to transfer the YCbCr space from frequency domain to space domain.
- Reconstruct the chrominance by upsampling and transfer the YCbCr space to RGB color space.
- Save the pixel value to the bitmap in RGB space and output the decompressed image.

In the following chapters, we will discuss the JPEG decoder in details and show the experimental results.

The JPEG Decoder

In this chapter, we will introduce the JPEG decoder step by step. Here we give an example of decoding a JPEG image "lena.jpg" compressed by Paint on Windows 7.

JPEG

In computing, JPEG - named after its creator the Joint Photographic Expert Group - (/'d?e?p?g/ JAY-peg) (seen most often with the. jpg extension) is a commonly used method of lossy compression for digital photography (i.e. images). The degree of compression can be adjusted, allowing a selectable tradeoff between storage size and image quality. JPEG typically achieves 10:1 compression with little perceptible loss in image quality, and is the file type most often produced in digital photography.

JPEG compression is used in a number of image file formats. JPEG/Exif is the most common image format used by digital cameras and other photographic image capture devices; along with JPEG/JFIF, it is the most common format for storing and transmitting photographic images on the World Wide Web. These format variations are often not distinguished, and are simply called JPEG. The term "JPEG" is an acronym for the Joint Photographic Experts Group, which created the standard. The MIME media type for JPEG is image/jpeg (defined in RFC 1341), except in Internet Explorer, which provides a MIME type of image/pjpeg when uploading JPEG images.

THE JPEG STANDARD

The name "JPEG" stands for Joint Photographic Experts Group, the name of the committee that created the JPEG standard and also other still picture coding standards. The "Joint" stood for ISO TC97 WG8 and CCITT SGVIII. In 1987 ISO TC 97 became ISO/IEC JTC1 and in 1992 CCITT became ITU-T. Currently on the JTC1 side JPEG is one of two sub-groups of ISO/IEC Joint Technical Committee 1, Subcommittee 29, Working Group 1 (ISO/IEC JTC 1/ SC 29/WG 1) – titled as Coding of still pictures. On the ITU-T side ITU-T SG16 is the respective body. The original JPEG group was organized in 1986, issuing the first JPEG standard in 1992, which was approved in September 1992 as ITU-T Recommendation T.81 and in 1994 as ISO/IEC 10918-1. The JPEG standard specifies the codec, which defines how an image is compressed into a stream of bytes and decompressed back into an image, but not the file format used to contain that stream. The Exif and JFIF standards define the commonly used file formats for interchange of JPEG-compressed images.

TYPICAL USAGE

The JPEG compression algorithm is at its best on photographs and paintings of realistic scenes with smooth variations of tone and color. For web usage, where the amount of data used for an image is important, JPEG is very popular. JPEG/Exif is also the most common format saved by digital cameras. On the other hand, JPEG may not be as well suited for line drawings and other textual or iconic graphics, where the sharp contrasts between adjacent pixels can cause noticeable artifacts. Such images may be better saved in a lossless graphics format such as TIFF, GIF, PNG, or a raw image format. The JPEG standard actually includes a lossless coding mode, but that mode is not supported in most products.

As the typical use of JPEG is a lossy compression method, which somewhat reduces the image fidelity, it should not be used in scenarios where the exact reproduction of the data is required (such as some scientific and medical imaging applications and certain technical image processing work). JPEG is also not well suited to files that will undergo multiple edits, as some image quality will usually be lost each time the image is decompressed and recompressed, particularly if the image is cropped or shifted, or if encoding parameters are changed – see digital generation loss for details. To avoid this, an image that is being modified or may be modified in the future can be saved in a lossless format, with a copy exported as JPEG for distribution.

JPEG COMPRESSION

JPEG uses a lossy form of compression based on the discrete cosine transform (DCT). This mathematical operation converts each frame/field of the video source from the spatial (2D) domain into the frequency domain (aka transform domain.) A perceptual model based loosely on the human psychovisual system discards high-frequency information, i.e. sharp transitions in intensity, and color hue. In the transform domain, the process of reducing information is called quantization. In simpler terms, quantization is a method for optimally reducing a large number scale (with different occurrences of each number) into a smaller one, and the transform-domain is a convenient representation of the image because the high-frequency coefficients, which contribute less to the over picture than other coefficients, are characteristically small-values with high compressibility. The quantized coefficients are then sequenced and losslessly packed into the output bitstream. Nearly all software implementations of JPEG permit user control over the compression-ratio (as well as other optional parameters), allowing the user to trade off picture-quality for smaller file size. In embedded applications (such as miniDV, which uses a similar DCT-compression scheme), the parameters are pre-selected and fixed for the application.

The compression method is usually lossy, meaning that some original image information is lost and cannot be restored, possibly affecting image quality. There is an optionallossless mode defined in the JPEG standard; however, this mode is not widely supported in products. There is also an interlaced "Progressive JPEG" format, in which data is compressed in multiple passes of progressively higher detail. This is ideal for large images that will be displayed while downloading over a slow connection, allowing a reasonable preview after receiving only a portion of the data. However, support for Progressive JPEGs is not universal; when Progressive JPEGs are received by programs that do not support them (such as versions of Internet Explorer before Windows 7) the software only displays the image after it has been completely downloaded. There are also many medical imaging and traffic systems that create and process 12-bit JPEG images, normally grayscale images. The 12-bit JPEG format has been part of the JPEG specification for some time, but again, this format is not as widely supported.

Lossless Editing

A number of alterations to a JPEG image can be performed losslessly (that is, without recompression and the associated quality loss) as long as the image size is a multiple of 1 MCU block (Minimum Coded Unit) (usually 16 pixels in both directions, for 4:2:0 chroma subsampling). Utilities that implement this include jpegtran, with user interface Jpegcrop, and the JPG_TRANSFORM plugin to IrfanView. Blocks can be rotated in 90 degree increments, flipped in the horizontal, vertical and diagonal axes and moved about in the image. Not all blocks from the original image need to be used in the modified one.

The top and left edge of a JPEG image must lie on a 8 × 8 pixel block boundary, but the bottom and right edge need not do so. This limits the possible lossless cropoperations, and also prevents flips and rotations of an image whose bottom or right edge does not lie on a block boundary for all channels (because the edge would end up on top or left, where – as aforementioned – a block boundary is obligatory). When using lossless cropping, if the bottom or right side of the crop region is not on a block boundary then the rest of the data from the partially used blocks will still be present in the cropped file and can be recovered. It is also possible to transform between baseline and progressive formats without any loss of quality, since the only difference is the order in which the coefficients are placed in the file.

Furthermore, several JPEG images can be losslessly joined together, as long as the edges coincide with block boundaries. jpeg supports 12-bit and 32-bit color as RGB.

JPEG FILES

The file format known as "JPEG Interchange Format" (JIF) is specified in Annex B of the standard. However, this "pure" file format is rarely used, primarily because of the difficulty of programming encoders and decoders that fully implement all aspects of the standard and because of certain shortcomings of the standard:

- Color space definition
- Component sub-sampling registration
- Pixel aspect ratio definition.

Several additional standards have evolved to address these issues. The first of these, released in 1992, was JPEG File Interchange Format (or JFIF), followed in recent years by Exchangeable image file format (Exif) and ICC color profiles. Both of these formats use the actual JIF byte layout, consisting of different markers, but in addition employ one of the JIF standard's extension points, namely the application markers: JFIF use APP0, while Exif use APP1. Within these segments of the file, that were left for future use in the JIF standard and aren't read by it, these standards add specific metadata. Thus, in some ways JFIF is a cutdown version of the JIF standard in that it specifies certain constraints (such as not allowing all the different encoding modes), while in other ways it is an extension of JIF due to the added metadata. The documentation for the original JFIF standard states:

JPEG File Interchange Format is a minimal file format which enables JPEG bitstreams to be exchanged between a wide variety of platforms and applications. This minimal format does not include any of the advanced features found in the TIFF JPEG specification or any application specific file format. Nor should it, for the only purpose of this simplified format is to allow the exchange of JPEG compressed images. Image files that employ JPEG compression are commonly called "JPEG files", and are stored in variants of the JIF image format. Most image capture devices (such as digital cameras) that output JPEG are actually creating files in the Exif format, the format that the camera industry has standardized on for metadata interchange. On the other hand, since the Exif standard does not allow color profiles, most image editing software stores JPEG in JFIF format, and also include the APP1 segment from the Exif file to include the metadata in an almost-compliant way; the JFIF standard is interpreted somewhat flexibly.

Strictly speaking, the JFIF and Exif standards are incompatible because they each specify that their marker segment (APP0 or APP1, respectively) appears first. In practice, most JPEG files contain a JFIF marker segment that precedes the Exif header. This allows older readers to correctly handle the older format JFIF segment, while newer readers also decode the following Exif segment, being less strict about requiring it to appear first.

JPEG Filename Extensions

The most common filename extensions for files employing JPEG compression are jpg and jpeg, though jpe, jfif and jif are also used. It is also possible for JPEG data to be embedded in other file types – TIFF encoded files often embed a JPEG image as a thumbnail of the main image; and MP3 files can contain a JPEG of cover art, in the ID3v2tag.

Color Profile

Many JPEG files embed an ICC color profile (color space). Commonly used color profiles include sRGB and Adobe RGB. Because these color spaces use a non-linear transformation, the dynamic range of an 8-bit JPEG file is about 11 stops; see gamma curve.

SYNTAX AND STRUCTURE

A JPEG image consists of a sequence of segments, each beginning with a marker, each of which begins with a 0xFF byte followed by a byte indicating what kind of marker it is. Some markers consist of just those two bytes; others are followed by two bytes indicating the length of marker-specific payload data that follows. (The length includes the two bytes for the length, but not the two bytes for the marker.) Some markers are followed by entropy-coded data; the length of such a marker does not include the entropy-coded data. Note that consecutive 0xFF bytes are used as fill bytes for padding purposes, although this fill byte padding should only ever take place for markers immediately following entropy-coded scan data (see JPEG specification section B.1.1.2 and E.1.2 for details; specifically "In all cases where markers are appended after the compressed data, optional 0xFF fill bytes may precede the marker").

JPEG CODEC EXAMPLE

Although a JPEG file can be encoded in various ways, most commonly it is done with JFIF encoding. The encoding process consists of several steps:

- The representation of the colors in the image is converted from RGB to Y'CBCR, consisting of one luma component (Y'), representing brightness, and two chromacomponents, (CB and CR), representing color. This step is sometimes skipped.

- The resolution of the chroma data is reduced, usually by a factor of 2. This reflects the fact that the eye is less sensitive to fine color details than to fine brightness details.
- The image is split into blocks of 8×8 pixels, and for each block, each of the Y, CB, and CR data undergoes the Discrete Cosine Transform (DCT), which was developed in 1974 by N. Ahmed, T. Natarajan and K. R. Rao; see Citation 1 in Discrete cosine transform. A DCT is similar to a Fourier transform in the sense that it produces a kind of spatial frequency spectrum.
- The amplitudes of the frequency components are quantized. Human vision is much more sensitive to small variations in color or brightness over large areas than to the strength of high-frequency brightness variations. Therefore, the magnitudes of the high-frequency components are stored with a lower accuracy than the low-frequency components. The quality setting of the encoder (for example 50 or 95 on a scale of 0–100 in the Independent JPEG Group's library) affects to what extent the resolution of each frequency component is reduced. If an excessively low quality setting is used, the high-frequency components are discarded altogether.
- The resulting data for all 8×8 blocks is further compressed with a lossless algorithm, a variant of Huffman encoding.

The decoding process reverses these steps, except the quantization because it is irreversible. In the remainder of this section, the encoding and decoding processes are described in more detail.

Encoding

Many of the options in the JPEG standard are not commonly used, and as mentioned above, most image software uses the simpler JFIF format when creating a JPEG file, which among other things specifies the encoding method. Here is a brief description of one of the more common methods of encoding when applied to an input that has 24bits per pixel (eight each of red, green, and blue). This particular option is a lossy data compression method.

Color Space Transformation

First, the image should be converted from RGB into a different color space called Y'CBCR (or, informally, YCbCr). It has three components Y', CB and CR: the Y' component represents the brightness of a pixel, and the CB and CR components represent the chrominance (split into blue and red components). This is basically the same color space as used by digital color television as well as digital video including video DVDs, and is similar to the way color is represented in analog PAL video and MAC (but not by analogNTSC, which uses the YIQ color space). The Y'CBCR color space conversion allows greater compression without a significant effect on perceptual image quality (or greater perceptual image quality for the same compression). The compression is more efficient because the brightness information, which is more important to the eventual perceptual quality of the image, is confined to a single channel. This more closely corresponds to the perception of color in the human visual system. The color transformation also improves compression by statistical decorrelation.

A particular conversion to Y'CBCR is specified in the JFIF standard, and should be performed for the resulting JPEG file to have maximum compatibility. However, some JPEG implementations in "highest quality" mode do not apply this step and instead keep the color information in the RGB color model, where the image is stored in separate channels for red, green and blue brightness components. This results in less efficient compression, and would not likely be used when file size is especially important.

Downsampling

Due to the densities of color- and brightness-sensitive receptors in the human eye, humans can see considerably more fine detail in the brightness of an image (the Y' component) than in the hue and color saturation of an image (the Cb and Cr components). Using this knowledge, encoders can be designed to compress images more efficiently.

The transformation into the Y'CBCR color model enables the next usual step, which is to reduce the spatial resolution of the Cb and Cr components (called "downsampling" or "chroma subsampling"). The ratios at which the downsampling is ordinarily done for JPEG images are 4:4:4 (no downsampling), 4:2:2 (reduction by a factor of 2 in the horizontal direction), or (most commonly) 4:2:0 (reduction by a factor of 2 in both the horizontal and vertical directions). For the rest of the compression process, Y', Cb and Cr are processed separately and in a very similar manner.

Block Splitting

After subsampling, each channel must be split into 8×8 blocks. Depending on chroma subsampling, this yields (Minimum Coded Unit) MCU blocks of size 8×8 (4:4:4 – no subsampling), 16×8 (4:2:2), or most commonly 16×16 (4:2:0). In video compression MCUs are called macroblocks.

If the data for a channel does not represent an integer number of blocks then the encoder must fill the remaining area of the incomplete blocks with some form of dummy data. Filling the edges with a fixed color (for example, black) can create ringing artifacts along the visible part of the border; repeating the edge pixels is a common technique that reduces (but does not necessarily completely eliminate) such artifacts, and more sophisticated border filling techniques can also be applied.

JPEG DECODING PROCESS

The JPEG decoder takes a JPEG image and converts it to a (uncompressed) bitmap using a decoding method called the baseline decoding process. A short introduction to the JPEG decoding process can be found in chapter 2 and appendix of the report "Design and implementation of a JPEG decoder" by Sander Stuijk. The multi-processor decoder is based on asingle-processor decoder that is in./c_prog/djpeg_orig. Figure 1 shows the process steps (blue) of the JPEG decoding process:

- VLD - Variable length decoding,
- ZZ - Zigzag scan,
- DQ - Dequantization,
- IDCT - Inverse discrete cosine transform,
- Color conversion (YUV to RGB) and reorder.

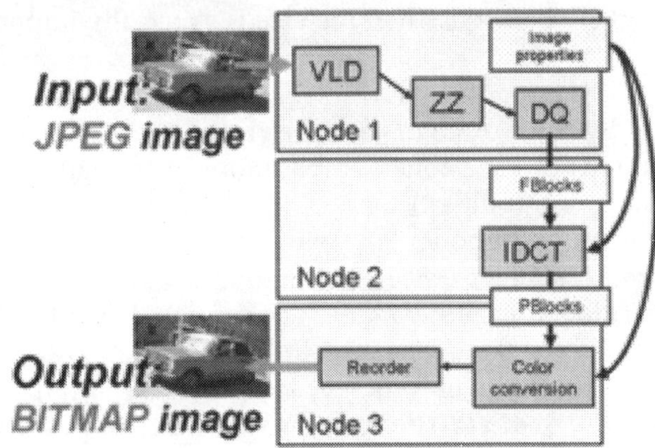

Fig. JPEG decoding process

The yellow blocks describe the data that is sent between nodes (mMIPS processors). See the next paragraph for more information.

DECODER PARTITIONING

The JPEG decoder process has been divided up into three nodes as depicted in figure 1. Each step runs on a separate node of the mMIPS network: node1 is at (X,Y) = (0,0), node2 is at (1,0), node3 is at (0,1) and the node (1,1) remains unused. This partition was chosen because a quick investigation of the sources of the original JPEG decoder revealed that partitioning just before and after IDCT-function was the easiest to realize. This choice also has the advantage that the Huffman decoding and dequantization tables required by the VLD and DQ units respectively do not need to be sent over the network. This partition leads to the distribution of the workload as shown in table.

Table. Workload for nodes 1, 2 and 3 in the JPEG Decoder.

Node	Steps + workload %	Total workload %
1	VLD (35%) + ZZ (5%) + DQ (10%)	50%
2	IDCT (20%)	20%
3	Color conversion (15%) + Reorder (15%)	30%

File structure

Table below describes the files and folders in./c_prog/djpeg_mmips.

Table. Overview of the files and folders in./c_prog/djpeg_mmips.

File	Contents / purpose
/dumps	Contains the files mips_ram.xXyY.dump, which give the data memory contents after a simulation has ended for nodes (X,Y). These files are moved to the folder automatically if the Linux shell script dolcc is used.
/example	This folder contains the simulation results of the JPEG decoder for an example image.
/node.xXyY	If the Linux shell script dogcc is used, then a node generates files named output_to_0xADDR.bin. The file contains the data the node tried to send to the node with relative address ADDR.
/test_images	Contains JPEG images that can be used to test the decoder.
color.*	Contains color_conversion() that converts YUV colors to their RGB equivalents.
djpeg.kdevprj djpeg.kdevses	Project files that describe the JPEG decoder in the Linux programming environment kdevelop.
dogcc	Use the script dogcc to compile and run the JPEG decoder from the Linux shell using gcc. It uses Makefile to compile and renames the output files in such a way that the communications pattern depicted in figure 1 is achieved. The script is for the 2-by-2 mMIPS NOC.
dolcc	Use the script dolcc X to compile the JPEG decoder using lcc and run a simulation for X minutes. The script is for the 2-by-2 mMIPS NOC.
dump	dump is used in the scripts dolcc to create and initialize the node data memories.
fast_int_idct.*	Inverse Discrete Cosine Transform (IDCT) using Integers.
huffman.*	Performs the Variable length decoding (VLD), Zigzag scan (ZZ) and Dequantization (DQ),.
jpeg.h	Constants, preprocessor symbols and data structures common to all nodes.
Makefile	Used by the script ./dogcc to compile the nodes using gcc.
mips	Ready to use hardware simulator for a 2x2 mMips NOC
mips_ram.empty.bin	Empty data memory which is used by the script dolcc to initialize the data memories.

parse.*	Contains mgetc() which allows functions within step1.c access to the input image. Also contains some other functions that involve retrieving specific data from the image.
stepX.*	Contains the function main() for node X. See figure 1 for a description of the JPEG decoding steps performed by each node.
strings.sh	Dumps the output of mprintf() for each node. Since the mMIPS NOC does not have a terminal, the printf() clone mprintf() is used to output to memory.
sunraster.*	Extracts a viewable bitmap image from the memory dump of the last node / step in the JPEG decoding process.
tree_vld.*	Creates the Huffman table.

Compile and Run

The sources of the multi-processor JPEG decoder are located in./c_prog/djpeg_mmips. The compilation steps needed to compile the JPEG decoder are comparable to those for gossip. The compilation process and run types for gossip are discussed in the application design flow. You need to make few modifications to the C sources and shell scripts if you want to change the input image of the JPEG decoder or change the memory layout of the mMIPSes.

Change the Input Image

A change of input image involves the following changes to the source code and scripts:

- Function load_image() in parse.c– The path and filename of the input image is hard coded in this function and needs to be changed.
- Header file jpeg.h– If the decoder is run on a PC/i386 directly (using the script dogcc), then the preprocessor symbols JPG_IMAGE_BUFFER_SIZE and BMP_IMAGE_BUFFER_SIZE determine the memory reserved to store the input and output images respectively.
- If the decoder is run on the NOC simulator (using the script dolcc) or the actual FPGA hardware, then the preprocessor symbol JPG_IMAGE_MAX_ADDR determines maximum address in the data memory that could contain input image data. An error is printed using mprintf() if the decoder tries to read beyond that address, but code execution continues. See themMIPS page for more information on its addressing modes and memory layout.
- The script dolcc–The JPEG image is copied to the start of mips_ram.x0y0.bin. This file contains the data memory of node 1.

Simulation Results

The subfolders dogcc and dolcc in the folder./c_prog/djpeg_mmips/

example contain the output of a run of dolcc and dogcc respectively for a 32x24 pixels color JPEG image of a surfer. Any output written to stdout or stderr during the execution of these scripts was saved in the file output_surfer.txt. For dolcc the dumps subfolder contains the contents of the data memories of the mMIPSes after completion. The output bitmap is in the data memory file mips_ram.x0y1.dump beginning at address 0x0 by default. khexedit (an X-Server like Hummingbird Exceed is needed) can be used to verify that it is exactly the same as the file output.ras that dogcc generates (see next).

The syntax of the HEX-editor is khexedit <filename>. For dogcc the node.xXyY subfolders contain any data that was sent and output.ras is the resulting bitmap (in Sun raster format). You can view the output file with eog (Eye of Gnome) if you have an X-Server like Hummingbird Exceed using the command: eog output.ras &

The decoding of surfer.jpg on a 2x2 mMIPS NOC using the hardware simulator took 29 hours on a Pentium III 1GHz processor running GNU/Linux 2.4.20 with 2048 MB of RAM. This process the took 604 milliseconds on the actual FPGA implementation. This simulation speed may be too low for some situations. The main two reasons for the low simulation speed are the cycle-accurate RTL level simulation and the lack of a multiply instruction. Fortunately, there are a number of things we can do to improve the performance of the simulator.

- Raise the abstraction level of the simulator from cycle-accurate RTL level simulation to non-synthesizable cycle-accurate or instruction set simulation.
- Increase the speed of the multiplication. The absence of a multiply instruction means that all multiplications need to be done using a software call (see the section soft ops on the mMIPS page). Any occurrence of the symbol '*' in C is translated into a call to the function __mul (the assembly implementation of this file is in./lcc/lib/emu_asm.s). This implementation is time-consuming and can be increased: multiplications can be inlined instead of invoked using a function call, negative operands in the multiplication can be inversed and operands could be sorted according to their value. Alternatively we could implement the multiplication in hardware. A faster simulator is one of the goals for the future.

READ HEADER

For the purpose of discussing the content of a JPEG file, we show the file in hexadecimal with PSPad which is a text editor. We can find that it comes with "FF DB" in the beginning, which is the marker "SOI". The meaning of the marker. "FF E0" comes after "FF D8", which is the beginning of the JFIF header. After the JFIF header, we will start to recode 6 important tables which are quantization table for luminance, quantization table for chrominance, Huffman table for luminance DC, Huffman table for luminance AC, Huffman table for chrominance DC and Huffman table for chrominance AC.

Table. Some markers of the JPEG header

Name	Code (HEX)	Description
SOI	FFD8	Start Of Image
EOI	FFD9	End Of Image
SOF	FFC0	Start Of Frame
SOS	FFDA	Start Of Scan
DHT	FFC4	Define Huffman Table
DQT	FFDB	Define Quantization Table
	FFE0	JFIF marker

The quantization tables are start from "FF DB".

For the quantization table – luminance:

- "00 43": The length of quantization table is 43-2("00 43")-1("00") =40(Hex) =64(Dec).
- "00": The marker of quantization table for luminance.
- "02 01"~"0C 0C": The content of quantization table for luminance, which should be record in zigzag.

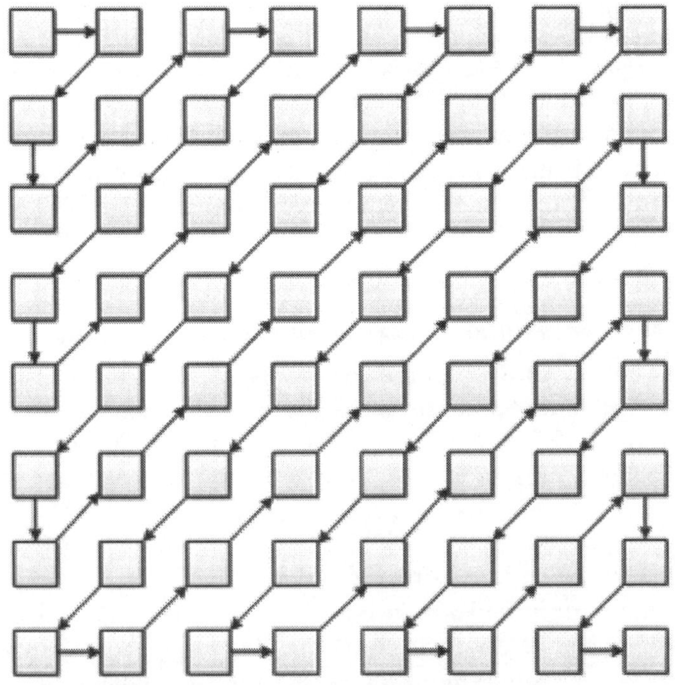

Fig.:Zigzag-scanning for 8x8 block.

Table. Quantization table for luminance

2	1	1	2	3	5	6	7
1	1	2	2	3	7	7	7
2	2	2	3	5	7	8	7
2	2	3	3	6	10	10	7
2	3	4	7	8	13	12	9
3	4	7	8	10	12	14	11
6	8	9	10	12	15	14	12
9	11	11	12	13	12	12	12

For the quantization table –chrominance:

- "00 43": The length of quantization table is 43-2("00 43")-1("00") =40(Hex) =64(Dec).
- "01": The marker of quantization table for chrominance.
- "02 02"~"0C 0C": The content of quantization table for chrominance, which should be record in zigzag.

Table. Quantization Table for Chrominance.

2	2	3	6	12	12	12	12
2	3	3	8	12	12	12	12
3	3	7	12	12	12	12	12
6	8	12	12	12	12	12	12
12	12	12	12	12	12	12	12
12	12	12	12	12	12	12	12
12	12	12	12	12	12	12	12
12	12	12	12	12	12	12	12

READ FRAME HEADER

- "FF C0" means start of the frame header which contains the information about the frame such as frame height, frame width, subsampling mode, etc. The following are the details of frame header:
- "00 11": The length of the frame header is 11(Hex) = 17(Dec).
- "08": Range of pixel value is 0~255(8 bit).
- "02 00": The height of the frame is 200(Hex) = 512(Dec).
- "02 00": The width of the frame is 200(Hex) = 512(Dec).
- "03": Number of components is 3 (Y, Cb, Cr).
- "01": ID of Y.
- "22": The horizontal sampling factor of Y is H_Y=2, and the vertical sampling factor of Y is V_Y=2.
- "00": Quantization table number of Y.

- "02": ID of Cb.
- "11": The horizontal sampling factor of Cb is H_{Cb}=1, and the vertical sampling factor of Cb is V_{Cb}=1.
- "01": Quantization table number of Cb.
- "03": ID of Cr.
- "11": The horizontal sampling factor of Cr is H_{Cr}=1, and the vertical sampling factor of Cr is V_{Cr}=1.
- "01": Quantization table number of Cr.

The frame sizes of each component can be calculated from frame height, frame width, and sampling factors. In this case, the maximum horizontal sampling factor is H_{max}=2 and maximum vertical sampling factor is V_{max}=2.

Thus, the frame sizes of each component are:

- Frame width of Y: $\text{frame width} \times \frac{H_Y}{H_{max}} = 512 \times \frac{2}{2} = 512$
- Frame height of Y: $\text{frame height} \times \frac{V_Y}{V_{max}} = 512 \times \frac{2}{2} = 512$
- Frame width of Cb: $\text{frame width} \times \frac{H_{Cb}}{H_{max}} = 512 \times \frac{1}{2} = 256$
- Frame height of Cb: $\text{frame height} \times \frac{V_{Cb}}{V_{max}} = 512 \times \frac{1}{2} = 256$
- Frame width of Cr: $\text{frame width} \times \frac{H_{Cr}}{H_{max}} = 512 \times \frac{1}{2} = 256$
- Frame height of Cr: $\text{frame height} \times \frac{V_{Cr}}{V_{max}} = 512 \times \frac{1}{2} = 256$

From the sampling factors, it can be found that Cb and Cr are downsampled in every 2 pixels in both vertical and horizontal direction. Therefore, the subsampling mode in this case is 4:2:0 as shown in Fig. below.

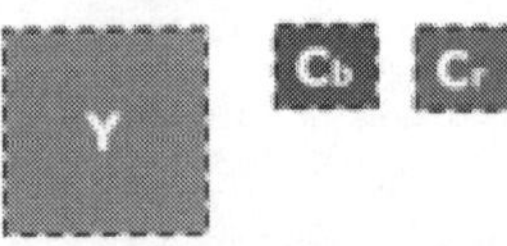

Read Huffman Table

In this section, we will read the Huffman table in the JPEG file first, and then cluster the Huffman table by the method Error! Reference source not found.

The following are the details of Huffman table for luminance DC:

- "FF C4": Define Huffman table which is the start marker.
- "00 1F": The number of category is 1F - 2("00 1F") - 1("00") – 10("00 01~00 00") =D(Hex)=12(Dec).
- "00": Huffman Table for luminance DC.
- "00 01"~"00 00": Numbers of codeword with codelength equal to 1~16. The detail of this case.
- "00 01"~"0A 0B": Category of Huffman table for luminance DC.

Now we can generate the Huffman table for luminance DC.

Table. Number of Codeword v.s. Codelength for Huffman Table for Luminance DC.

Codelength	1	2	3	4	5	6	7	8	9	10	11	12	13	14	15	16
Number of codeword	0	1	5	1	1	1	1	1	1	0	0	0	0	0	0	0

Algorithm 1: Generate Huffman table

```
        i = 0, codevalue = 0;
for(k = 1; k <= 16; k++)
                        {
                          for(j = 1; j <= number_of_codeword[k]; j++)
                          {
                          codeword[i] = codevalue;
                          codelength[i] = k;
                          codevalue++;
                          i++;
                          }
                          codevalue *= 2;
                        }
```

Table. Huffman Table for luminance DC.

Category	Codelength	Codeword
00	2	00
01	3	10
02	3	11
03	3	100
04	3	101
05	3	110
06	4	1110
07	5	11110
08	6	111110
09	7	1111110
0A	8	11111110
0B	9	111111110

The following are the details of Huffman table for luminance AC:

- "FF C4": Define Huffman table which is the start marker.
- "00 B5": The number of category is B5 - 2("00 B5") - 1("00") – 10("00 02~00 7D") =A2(Hex)=162(Dec).
- "10": Huffman Table for luminance AC.
- "00 02"~"00 7D": Numbers of codeword with codelength equal to 1~16.
- "01 02"~"F9 FA": Category of Huffman table for luminance AC.

Now we can generate the Huffman table for luminance AC.

Table. Number of codeword v.s. Codelength for Huffman table for Luminance AC.

Codelength	1	2	3	4	5	6	7	8	9	10	11	12	13	14	15	16
Number of codeword	0	2	1	3	3	2	4	3	5	5	4	4	0	0	1	125

Table. Part of Huffman Table for Luminance AC.

Category	Codelength	Codeword
01	2	00
02	2	01
03	3	100
00	4	1010
04	4	1011
11	4	1100
05	5	11010
12	5	11011
21	5	11100
31	6	111010
41	6	111011
…	…	…
EA	16	1111111111110100
F1	16	1111111111110101
F2	16	1111111111110110
F3	16	1111111111110111
F4	16	1111111111111000
F5	16	1111111111111001
F6	16	1111111111111010
F7	16	1111111111111011
F8	16	1111111111111100
F9	16	1111111111111101
FA	16	1111111111111110

The following are the details of Huffman table for chrominance DC:

- "FF C4": Define Huffman table which is the start marker.
- "00 1F": The number of category is 1F - 2("00 1F") - 1("00") – 10("00 01~00 00") =D(Hex)=12(Dec).
- "01": Huffman Table for chrominance DC.
- "00 03"~"00 00": Numbers of codeword with codelength equal to 1~16.
- "00 01"~"0A 0B": Category of Huffman table for chrominance DC.

Now we can generate the Huffman table for chrominance DC.

Table. Number of Codeword v.s. Codelength for

Codelength	1	2	3	4	5	6	7	8	9	10	11	12	13	14	15	16
Number of codeword	0	3	1	1	1	1	1	1	1	1	1	0	0	0	0	0

Table. Huffman table for chrominance DC

Category	Codelength	Codeword
00	2	00
01	2	01
02	2	10
03	3	110
04	4	1110
05	5	11110
06	6	111110
07	7	1111110
08	8	11111110
09	9	111111110
0A	10	1111111110
0B	11	11111111110

The following are the details of Huffman table for chrominance AC:

- "FF C4": Define Huffman table which is the start marker.
- "00 B5": The number of category is B5 - 2("00 B5") - 1("00") – 10("00 02~00 7D") =A2(Hex)=162(Dec).
- "11": Huffman Table for chrominance AC.
- "00 02"~"02 77": Numbers of codeword with codelength equal to 1~16.
- "00 01"~"F9 FA": Category of Huffman table for chrominance AC.

Now we can generate the Huffman table for chrominance AC.

Table. Number of codeword v.s. Codelength for Huffman Table for Chrominance AC.

Codelength	1	2	3	4	5	6	7	8	9	10	11	12	13	14	15	16
Number of codeword	0	2	1	2	4	4	3	4	7	5	4	4	0	1	2	119

Table. Part of Huffman table for chrominance AC.

Category	Codelength	Codeword
00	2	00
01	2	01
02	3	100
03	4	1010
11	4	1011
04	5	11000
05	5	11001
21	5	11010
31	5	11011
06	6	111000
12	6	111001
...	...	...
E9	16	1111111111110100
EA	16	1111111111110101
F2	16	1111111111110110
F3	16	1111111111110111
F4	16	1111111111111000
F5	16	1111111111111001
F6	16	1111111111111010
F7	16	1111111111111011
F8	16	1111111111111100
F9	16	1111111111111101
FA	16	1111111111111110

PEER-TO-PEER MULTICASTING INSPIRED BY HUFFMAN CODING

For previous few years peer-to-peer (P2P) related flows (overlay traffic) have represented a vast majority of the whole Internet traffic. Thus, P2P networking has become an important branch of today's telecommunications. It involves not only file-sharing (BitTorrent, Ares Galaxy, Gnutella), but also multimedia streaming (PPLive, SopCast), Internet telephony (Skype), anonymous routing (Tor), and many aspects of content distribution networks or cloud networking.

With the increased interest in quality-requiring applications, the problem of the quality of service (QoS) assurance is a real challenge in P2P-based streaming. Live streaming transfers based on the P2P paradigm impose strict constraints on latency between a video server and peers (end-users). Since the requirements are not fully addressed by the characteristics of P2P networking, a considerable attention has been put on the overcoming techniques. One of the most basic parameters that highly influence the quality experienced by end-users is the effective downloading speed and its stability over time. A buffering process, observed by end-users as a transient interruption of content delivery, can be minimized, or even avoided, when oscillations of the downloading speed are small. To accomplish the gapless playback, the P2P application-layer links should be set over short distances with additional care about the available capacity. This task is a nontrivial problem as P2P systems were invented to be underlay-agnostic. In practice, they have not usually recognized proximity between the candidate peers nor estimate attainable bandwidth. This situation has changed and the focus has been brought to the so-called localization-aware overlays.

Here, we elaborate on the localization-aware P2P overlays: an organization of the transmission tree is not spontaneous (following the flowing process of the peers to the overlay), but instead it uses the underlying information on the Internet paths. In effect, the video quality is improved. Our method is called "Huffmies" due to its inspiration by the Huffman source coding. The Huffman algorithm constructs a tree representing a code of the shortest average codeword length, calculated as the sum of the paths from the root to the leaves (messages). The elements of the sum are weighted by the probabilities associated with the leaves. In Huffmies, a similar mechanism is used to create a multicast streaming tree, where the root is a video server and leaves represent the peers. We aim at generating the shortest average path from the root to a leaf, while the weight of a leaf is related to a measure based on the underlay-related knowledge, namely, round-trip-time. With the relationship of our algorithm to the Huffman method providing the optimal tree, we are able to show the optimality of Huffmies.

BACKGROUND AND PREVIOUS WORKS

With regard to the routing topology, there are the following three types of P2P streaming systems used in practice:(a)push: structured, based mainly on multicasting tree(s);(b)pull: unstructured, similar to the mesh-based file-sharing BitTorrent system;(c)hybrid: combined push-pull, based on a two-layer or two-phased mesh-tree architecture.Currently, the majority of existing implementations and methods are related to the second type. This stems from their flexibility and the high fault tolerance. However, we focus on the tree-based multicasting. It has also received strong research attention due to its control potential. Additionally, by the usage of the tree-like approach, we are able to prove the bound optimality with the inspiration of the Huffman source-coding algorithm. Additionally, we study the reliability performance of our method to prove its usefulness in the case of random failures.

The challenge related to P2P networks is to find the optimal topology able to satisfy each of the three entities involved in the overlay content delivery: (a) peers, (b) the Internet service providers (ISPs), and (c) content providers. Each of them has its own requirements and prominent objectives. A peer requires the high streaming rate of the content, short start-up times, and lack of prefetching. In particular, in live video transmissions, small delays between the content source and end-users are of vital importance. While interdomain traffic load is essential to ISPs, a reduction of that load should not affect the quality experienced by customers. Finally, goals of a content provider are in-between the end-users and ISPs' needs. The content provider appreciates a high customer satisfaction (enhanced video quality) and favorable network resource utilization, as it enables a successful cooperation with operators. Our method pays to the end-user QoS satisfaction; however it reconciles also other entities' matters which are respected and considered in the simulation study.

A lot of effort has been put to improve the operation of P2P systems by taking into account the network constraints; for review see. Sometimes, even a strong cooperation between an overlay operator and a network carrier is assumed. The topic has been recognized as an important issue for the Internet community. The research was mainly focused on file-sharing with introduction of methods like the so-called biased neighbor selection in the BitTorrent context. They rely on centrally managed devices impeding dynamic and quick decisions in case of nodes departure or network partitions. Therefore, their application prospects in video multicast networks are limited, especially when failures are considered.

A classification of approaches using locality awareness in P2P overlays can be found in (a) awareness of the ISPs network, to which a peer is assigned to; (b) storage of some additional information on other peers, for example, when they serve as supernodes; (c) delay assessments; (d) geolocation information as a base for the peer node selection. Our research follows the third avenue; that is, we base on the latency measured by. This group of approach is quite well represented. The streaming tree construction idea presented in involved just a delay-aware technique to build a network friendly tree (NFT). In this scheme, an arriving peer contacts and requests a streaming flow from the node that has the smallest among all peers in the overlay swarm. In contrast to our approach, the tree is constructed in a receiver-centric way and it does not adapt to transient network conditions. A similar concept of grouping peers into clustered trees. However, this method assumes evaluating between an end-user and static landmarks (like DNS servers), instead of ordinary overlay peers. Our approach is based on the between peers.

On the other hand, a proposal of Kovacevic et al. was based on the geolocation system, introducing zones and a two-layered, distance-aware system to accomplish strict QoS requirements. We do not assume such a sophisticated method, but base on the "application-layer ping," which is very simple to implement. The authors of proposed to construct an optimized multicasting P2P tree, assuming that the coordination data is given in advance. Again, we decided to use a much easier method.

Overview of Huffman Coding

A Huffman code is the minimum redundancy source code, where each message (out of messages) is represented as a prefix-free codeword (a "message code")., a codeword length, equal to the number of symbols in codeword, depends on an occurrence probability of the th message generation, so that the average message lengthis minimized. Each codeword is constructed using symbols from an alphabet of letters (the -ary alphabet). We skip the details of the code construction, but we remind that the ultimate idea is based on the fact that messages with higher are represented with shorter codewords. A Huffman ensemble code, that is an agreement between the source encoder and the decoder about corresponding messages and codewords, can be presented in the tree form.

Thus, an ensemble code and its tree are equivalent. In such a labeled tree, called here a Huffman tree, the internal and leaf nodes have different meanings. The leaf corresponds to a single message, labeled with its occurrence probability. The internal node is just an auxiliary vertex with labels equal to the sum of its children's label values. In tree, the edges are also uniquely labeled with one of symbols. Edge labels over a path from the tree root to a given leaf define a codeword representing the respective message. Such a definition of tree is used also in to create an adaptive Huffman algorithm. The original static (two-pass) and newer adaptive (one-pass) Huffman algorithms differ in the ensemble code construction method and its exchange between the encoder and decoder. However, both ensemble codes have the same structure and the value. Therefore, for the sake of interpretation purposes, we can use the static and adaptive algorithms interchangeably, since results for those approaches are the same.

Huffmies Design

Below, we present "Huffmies," the centralized and distributed approaches that create a streaming topology for a multicast tree-enabled P2P overlay network. Both approaches are inspired by the Huffman algorithm.

Centralized Huffmies

Each P2P streaming network uses a content server, which can upload to only a limited number of nodes simultaneously due to the bandwidth and computing resource limits. Among all the active peers subscribed to the stream, the server has to choose at most peers that will receive the data directly from it. The centralized Huffmies algorithm uses a single device managing all connections between the peer nodes in the entire overlay swarm. For the rest of this subsection we assume that server, which is a reliable, rich in resources, and credible node, can take responsibility of the overlay network construction and control. Hence, the terms "centralized device" and "server" will be used interchangeably here.

The Huffmies method introduces value, namely, peer aggravation. It forms a concise representation of peer 's network resources in such a way that higher values indicate inferior resources due to a tight bandwidth limit or high latency of the network access. Finding strict and accurate value of is not a trivial problem, since it is influenced by many unpredictable parameters (the location, available capacity, congestion). We simply use the round-trip-time metric () between a peer and the server or between a peer and another peer. The high value suggests that the given peer has considerably limited network resources or can be located in a long distance from the source node (or another peer). Nonetheless, more complex and sophisticated metrics can be involved here as well. Furthermore, as a peer receives data using a multiple-hop path, the sum of all values over this path emphasizes service aggravation of the connection.

In the centralized scheme, only the streaming server gathers all individual values, independently measured by the peers and sent to the server. The final value of is defined as an aggregated sum of s between peer and all other peers in the swarm, including the server:where represents the evaluated to node observed from the peer 's point of view (we set). All probing packets are analyzed individually by each node in a swarm and the computed result of is forwarded to the server.

As the central device has the entire map, it builds aHuffman tree, analogously to the Huffman coding with Vitter's algorithm extension, in a manner that is treated like a probability of message generation. In a transmission tree we are constructing, a codeword length is defined as a path length from the root (the streaming server) to leaf:Let us remind that the Huffman tree is a form, where only leaves represent something real (i.e., the messages). Therefore, in order to transform a Huffman tree to a transmission tree, a process calledclimbing is introduced. It consists in using the internal vertices as something of real meaning, that is, as the peers that are indeed present in the multicast tree to both download the data and simultaneously upload it to other peers. The climbing starts from the most bottom layer where each internal node, noted as "X", is replaced by the node with the lowest value beneath this auxiliary node.

While going upward, it can happen that the minimum value of characterizes a node that already has downstream nodes. In such a case, peer 's child with a minimum peer aggravation has to take its place and the node is free to jump into an upper layer. Such a process creates atransmission tree with a bounding number of children equal to.

Hence, parameter is considered as a maximum number of simultaneous unicast connections that a given peer can serve simultaneously. To simplify, we assume the same value for all peers. However, a dynamic adjustment of this value, for instance depending on the swarm size, streaming ratio, or peer's resources, is also possible.

Optimality of the Centralized Huffmies

In general, P2P tree-based streaming dissemination has to cope with conflicting objectives to design either a narrow or a short transmission tree. A wide tree, where each peer has to send data to numerous neighbors, divides the peer's upload capacity among all children. This can inflict the degradation of quality of experience (QoE), that is, the perceived quality, for the downstream peers. On the other hand, a high tree involves long paths between the streaming source and the peers. Such a tree should be averted since it imposes additional delays and decreases the system reliability in case of an intermediate node fault or departure. Thus, a convenient trade-off between the width and height should be found. The Huffmies method involves optimization of the worst-case peer aggravation suffered by an entire overlay swarm.

In general, each peer that arises on a path from the streaming server to a single destination node introduces bandwidth and delay limits that inflict a higher probability of service quality degradation for downstream nodes. Sum of all intermediate nodes' on a path from the server to a destination node is called a service aggravation. We model a deteriorated QoE caused by a multihop overlay path and observed by peer, as, wherewhere is a path from the server to the destination peer in a transmission tree. The main goal of the Huffmies method is to minimize this parameter for the entire overlay swarm.

We can note the two following fundamental facts. Theorem: The Huffmies algorithm presented in Section 4.1 ensures that each peer receives the data from a node that has a smaller or equal peer aggravation value. Proof. We show it by a contradiction. Assuming that peer receives data from node such that, leads to a contradictory statement, since peer, as a peer with smaller value among other siblings, would be elected for climbing instead of. Theorem: An upper bound for service aggravation suffered by peer in the centralized

Huffmies is. Proof. From the fact that the climbing process performed on the Huffman tree, as described in Section4.1, only uplifts some nodes into higher layers, then is an upper bound for the length of the transmission path from the server to the peer. On the basis of Theorem 1 we also state that bounds the maximum peer aggravation of nodes that can occur over this path. That provides the upper bound. According to the Huffman coding algorithm, which minimizes average message length (given in (1)), and Theorem 2 we can see that the centralized Huffmies algorithm minimizes the service aggravationbound for all nodes in the overlay swarm.

Distributed Huffmies

Application of Huffmies in a huge swarm inflicts a ping packet flood, since each peer has to evaluate between all other network members. Indeed, a number of such probes at the order of limit the practicality of the centralized scheme. Overview of the consecutive stages in the distributed Huffmies algorithm. The principal difference between the distributed and centralized version consists in a subjective estimation of the value accomplished by a single node in a P2P network. In the first step, the server estimates to each member in a swarm, which is assumed as its. Then, on the basis of peers' values, the server forms disjoint groups of peers. Among all the members of each group, the server elects a single peer to which it will upload the stream directly. Such a peer is called the group leader and it has to replicate the data towards the rest of the group nodes called an offspring of a given leader. The decision criteria used by the leaders to disseminate the data are completely independent and the server does not have any influence or authority regarding this issue. Practically, a leader performs the same steps as those of the server in the first step; that is, it selects the best leaders among its offspring and assigns them a next-level offspring (i.e., offspring of those nodes). Aleader bases its decision on independently obtained values. In the next steps, such a procedure is consecutively repeated until a pool of all peers in the overlay network is exhausted.

The key point, where the Huffman algorithm is engaged, concerns the selection of the leader nodes and assigning them appropriate peers to disseminate data downstream. To accomplish this task, each leader assigns a codeword for all of peers of its offspring, analogously to the centralized approach. Thus, it is necessary to calculate the Huffman tree in each group. Note that in this case, contrary to the centralized Huffmies, is assessed only from a group leader (or the server) to each peer, but not between all pairs of peers. The corresponding probabilities, are equal to evaluated over the individual probe between the leader and peer. After applying a codeword for each peer in, all peers whose first digit is the same constitute one group, obviously disjoint with other groups. Among all members of a group, a peer with the lowest value is selected as a group leader. Now, this leader is fully responsible for dissemination of data towards other members of its group.

Looking at this issue from the Huffman tree viewpoint, a group is constituted by all peers, whose corresponding leaves belong to the same root's subtree.

Such a distributed algorithm significantly diminishes aggregated number of links that are engaged in the probing process. While the streaming server has to evaluate all ??pings to all nodes in the swarm, other peers in lower transmission tree layers make a decision about much smaller groups, whose cardinality deteriorates at the order of, where is a peer's layer index in a transmission tree.

The Huffmies method clusters peers into groups, so that the sum of members' aggravation is almost equal for all groups. That means that peers with large values of will be present in groups of smaller number of members. This is a very advantageous and practical property. Let us remind that a low value for a given peer means that it inflicts a small negative effect on other children. We argue that it is highly recommended to create groups with a comparable sum of aggravation indicators. For instance, one group consisting of peers that are fast, reliable, and localized in a short distance from the source (all those parameters result in low) can aggregate a higher number of peers and the transmission quality will not be reduced. On the contrary, a high value of a member in a group prevents from accepting too many other peers. It prevents from low QoS parameters (like high delay and low streaming rate), experienced by many peers, when peer (with poor resources) becomes a relay point. In particular, when a peer has a notably higher (with comparison to the values of for other peers), caused by low-quality transmission (e.g., wireless access), it will constitute an independent group receiving data directly from the server to alleviate the negative effects related to this peer's poor resources.

Churn Effect

The considerations presented above assume the static environment: all peers are permanently on-line. However, in a real overlay, peers can freely join and depart the P2P network. In result, the population of the swarm and the connections within it often change, what is known as the churn effect. To avoid frequent chaotic rearrangements of connections in case of every peer joining or departure, we assume that the overlay members apply the Huffmies algorithm only in a specified point of time, calledreorganization. In the interim, every new peer is temporarily serviced by a random node. On the other hand, for each leader of a group, a backup leader is defined. It is selected as a peer with the second lowest value in the group. The backup leader takes care of a group in case of the leader departure. Therefore, a single peer departure does not destroy the existing multicast tree.

READ BITSTREAM OF HUFFMAN CODING

"FF DA" means start of scan which is the start marker of the bitstream of Huffman code, and FF D9" is the end marker of the bitstream, which means "end of image". The remains between "FF DA" and "FF D9"are the Huffman codes that we need to decode.

EXPERIMENTAL RESULTS AND CONCLUSION

In this chapter, we will show our experimental results. The following are the demo steps of our program, compiled on Windows 7 64bit with Visual Studio 2010. JPEGDecoderDemo.exe is under the folder: JPEGDecoderDemo\JPEGDecoder Demo\Release\.

HOW TO RUN THIS PROGRAM

Be sure that *"JPEGDecoderDemo.exe"* and JPEG file in the same folder, then double click *"JPEGDecoderDemo.exe"* and enter the input JPEG filename and output bmp filename. If choosing "Yes" for the question "Creat 'Parser.txt'?", it will create "Parser.txt" in the same folder, which shows the information such as quantization tables, Huffman tables, and decoded blocks YCbCr.

```
*************************** Quantization Table ***************
quantization_table_luminance
     8       6       5       8      12      20      26      31
     6       6       7      10      13      29      30      28
     7       7       8      12      20      29      35      28
     7       9      11      15      26      44      40      31
     9      11      19      28      34      55      52      39
    12      18      28      32      41      52      57      46
    25      32      39      44      52      61      60      51
    36      46      48      49      56      50      52      50

quantization_table_chrominance
     9       9      12      24      50      50      50      50
     9      11      13      33      50      50      50      50
    12      13      28      50      50      50      50      50
    24      33      50      50      50      50      50      50
    50      50      50      50      50      50      50      50
    50      50      50      50      50      50      50      50
    50      50      50      50      50      50      50      50
    50      50      50      50      50      50      50      50

***************************** Huffman Table ****************
---------Huffman Table luminance DC---------
Category  Codelength  Codeword
   00         2         00
   01         3         010
   02         3         011
   03         3         100
   04         3         101
   05         3         110
   06         4         1110
   07         5         11110
   08         6         111110
   09         7         1111110
   0A         8         11111110
   0B         9         111111110
```

```
---------Huffman Table luminance AC---------
Category  Codelength  Codeword
  01         2          00
  02         2          01
  03         3          100
  00         4          1010
  04         4          1011
  11         4          1100
  05         5          11010
  12         5          11011
  21         5          11100
```

Fig. Parts of lena.jpg in Parser.txt.

```
---------------------------------------------------------------Decode Y--
Before DeQuantize
  -20,   -2,   -1,    1,    0,    1,    0,   -1,
   -7,    1,    0,    1,    1,    0,    0,    1,
   -4,    1,   -1,   -1,    1,    0,    0,    0,
    1,    0,    1,    0,    0,    0,    0,    0,
    1,   -1,    0,    0,    0,    0,    0,    0,
   -1,   -1,    0,    0,    0,    0,    0,    0,
    0,    0,    0,    0,    0,    0,    0,    0,
    0,    0,    0,    0,    0,    0,    0,    0,

Before IDCT
 -160,  -12,   -5,    8,    0,   20,    0,  -31,
  -42,    6,    0,   10,   13,    0,    0,   28,
  -28,    7,   -8,  -12,   20,    0,    0,    0,
    7,    0,   11,    0,    0,    0,    0,    0,
    9,  -11,    0,    0,    0,    0,    0,    0,
  -12,  -18,    0,    0,    0,    0,    0,    0,
    0,    0,    0,    0,    0,    0,    0,    0,
    0,    0,    0,    0,    0,    0,    0,    0,

After IDCT
  -26,  -44,  -37,  -25,  -25,  -35,  -29,  -26,
  -16,  -30,  -28,  -20,  -25,  -32,  -32,  -36,
  -24,  -30,  -29,  -20,  -25,  -19,  -22,  -32,
  -25,  -25,  -28,  -18,  -23,   -3,   -6,  -18,
  -10,   -8,  -18,   -9,  -24,    2,   -3,  -13,
  -13,   -9,  -21,   -7,  -30,   -3,  -11,  -14,
  -27,  -20,  -28,   -4,  -32,   -8,  -17,  -13,
  -25,  -16,  -22,    7,  -27,   -8,  -22,  -15,


4 Y blocks are decoded. now
------------------------------------------------------------------Decode Cb-
Before DeQuantize
  -12,    1,    1,    0,    0,    0,    0,    0,
    2,    0,    0,    0,    0,    0,    0,    0,
    0,    0,    0,    0,    0,    0,    0,    0,
    0,    0,    0,    0,    0,    0,    0,    0,
    0,    0,    0,    0,    0,    0,    0,    0,
    0,    0,    0,    0,    0,    0,    0,    0,
    0,    0,    0,    0,    0,    0,    0,    0,
    0,    0,    0,    0,    0,    0,    0,    0,

Before IDCT
 -108,    9,   12,    0,    0,    0,    0,    0,
   18,    0,    0,    0,    0,    0,    0,    0,
    0,    0,    0,    0,    0,    0,    0,    0,
    0,    0,    0,    0,    0,    0,    0,    0,
    0,    0,    0,    0,    0,    0,    0,    0,
```

Fig. Parts of lena.jpg in Parser.txt.

If choosing "Yes" for the question "Display the details of jumping from cluster to cluster?", then it will display the details of decoding (bitstream in buffer, cluster, decoded symbols, decoded DC terms and AC terms, etc.) step by step in command window.

```
--------------------------------Reading bitstream -------------------------
1100:received bits in the buffer
buffer = 12
Now we are in the Cluster <a>
current_content=28
The content of the table at this location is 0/28, as the sign/magnitude
Decoded Symbol: Run/Size 1/1

ACterm = 1


--------------------------------Reading bitstream -------------------------
1100:received bits in the buffer
buffer = 12
Now we are in the Cluster <a>
current_content=28
The content of the table at this location is 0/28, as the sign/magnitude
Decoded Symbol: Run/Size 1/1

ACterm = 1


--------------------------------Reading bitstream -------------------------
1110:received bits in the buffer
buffer = 14
Now we are in the Cluster <a>
current_content=-2
The content of the table at this location is 1/2, as the sign/magnitude.
The symbol is not in cluster <a> and refer to location 2 in the ST,
assigned to cluster <c>.

00:received bits in the buffer
buffer = 0
Now we are in the Cluster <c>
current_content=2
The content of the table at this location is 0/2, as the sign/magnitude.
Decoded Symbol: Run/Size 2/1
```

Fig. Details of decoding steps in command window.

3

Powering Servos

The Parallax Homework Board is suggested as the platform for this module. As you begin to use the servos it will be important to have a second power supply to power the servos. The project will work best if the 9 v. battery is used for logic control and the 4 pack AA 6 v. supply is attached directly to the servos and not to the Vin or Vdd. The photo shown here indicates how to attach the power line from the battery. Notice that the white painted wire (positive) is attached directly to the breadboard pins attached to the red wire on the servos. The black, non painted wire is plugged into the Vss header block.This is an important step. If the white stripped wire is placed into the Vin or Vdd while a 9v. battery is attached to the main board, there will be overload issues with the batteries.

WHAT IS A SERVO MOTOR

Servo motors (or servos) are self-contained electric devices (see Figure 1 below) that rotate or push parts of a machine with great precision. Servos are found in many places: from toys to home electronics to cars and airplanes. If you have a radio-controlled model car, airplane, or helicopter, you are using at least a few servos. In a model car or aircraft, servos move levers back and forth to control steering or adjust wing surfaces. By rotating a shaft connected to the engine throttle, a servo regulates the speed of a fuel-powered car or aircraft. Servos also appear behind the scenes in devices we use every day.

Electronic devices such as DVD and Blu-ray Disc players use servos to extend or retract the disc trays. In 21st-century automobiles, servos manage the car's speed: The gas pedal, similar to the volume control on a radio, sends an electrical signal that tells the car's computer how far down it is pressed. The car's computer calculates that information and other data from other sensors and sends a signal to the servo attached to the throttle to adjust the engine speed. Commercial aircraft use servos and a related hydraulic technology to push and pull just about everything in the plane.

Fig. This assortment of servos is available in stores and by mail order. Servos range in price and application.

And of course, robots might not exist without servos. You see servo-controlled robots in almost every movie (those complex animatronic puppets have dozens of servos), and you have probably seen a number of robotic animal toys for sale. Smaller laboratory robots also use servos to move their joints. Hobby servos come in a variety of shapes and sizes for different applications. You may want a large, powerful one for moving the arm of a big robot, or a tiny one to make a robot's eyebrows go up and down. Two sizes you can find in a hobby store— an inexpensive common size and a more expensive miniature one.

Fig. Two common servo sizes. The standard servo on the left can range in power or speed to move something quickly, or it can accommodate a heavier load, such as steering a big radio-controlled monster truck or lifting the blade on a radio-controlled earthmover toy. The miniature servo is about the size of a U.S. quarter and is intended for applications where smallness is a critical factor but a lot of power is not.

HOW DOES A SERVO MOTOR WORK?

The simplicity of a servo is among the features that make them so reliable. The heart of a servo is a small direct current (DC) motor, similar to what you might find in an inexpensive toy. These motors run on electricity from a battery and spin at high RPM (rotations per minute) but put out very low torque (a twisting force used to do work— you apply torque when you open a jar). An arrangement of gears takes the high speed of the motor and slows it down while at the same time increasing the torque. (Basic law of physics: work = force x distance.) A tiny electric motor does not have much torque, but it can spin really fast (small force, big distance). The gear design inside the servo case converts the output to a much slower rotation speed but with more torque (big force, little distance). The amount of actual work is the same, just more useful. Gears in an inexpensive servo motor are generally made of plastic to keep it lighter and less costly. On a servo designed to provide more torque for heavier work, the gears are made of metal (see Figure 4 below) and are harder to damage.

Fig. The gears in a typical standard-size servo are made of plastic and convert the fast, low-power motion of the motor (on the right) to the output shaft (on the left).

Fig. In a high-power servo, the plastic gears are replaced by metal ones for strength. The motor is usually more powerful than in a low-cost servo and the overall output torque can be as much as 20 times higher than a cheaper plastic one. Better quality is more expensive, and high-output servos can cost two or three times as much as standard ones.

With a small DC motor, you apply power from a battery, and the motor spins. Unlike a simple DC motor, however, a servo's spinning motor shaft is slowed way down with gears. A positional sensor on the final gear is connected to a small circuit board. The sensor tells this circuit board how far the servo output shaft has rotated. The electronic input signal from the computer or the radio in a remote-controlled vehicle also feeds into that circuit board. The electronics on the circuit board decode the signals to determine how far the user wants the servo to rotate. It then compares the desired position to the actual position and decides which direction to rotate the shaft so it gets to the desired position.

Fig. The circuit board and DC motor in a high-power servo. Did you notice how few parts are on the circuit board? Servos have evolved to a very efficient design over many years.

Imagine you are playing catch with a friend on a sports field. You stand at one end and want your friend to go out for a long throw. You could keep calling out "farther, farther, farther" until she got as far away as you wanted. But if she went out farther than you can throw, you would have to call out "closer" until she got back to the right spot. If she were a simple motor in a robot arm and you were the microprocessor, you would have to spend some of your time watching what she did and giving her commands to move her back to the right spot (this is called a feedback loop). If she were a servo motor, you could just say "go out exactly 4.5 meters" and know that she would find the right spot. That is what makes servo motors so useful: once you tell them what you want done, they do the job without your help. This automatic seeking behavior of servo motors makes them perfect for many robotic applications.

TYPES OF SERVO MOTORS

Servos come in many sizes and in three basic types: positional rotation, continuous rotation, and linear.

- Positional rotation servo: This is the most common type of servo motor. The output shaft rotates in about half of a circle, or 180 degrees. It has physical stops placed in the gear mechanism to prevent turning beyond these limits to protect the rotational sensor. These common servos are found in radio-controlled cars and water- and aircraft, toys, robots, and many other applications.

- Continuous rotation servo: This is quite similar to the common positional rotation servo motor, except it can turn in either direction indefinitely. The control signal, rather than setting the static position of the servo, is interpreted as the direction and speed of rotation. The range of possible commands causes the servo to rotate clockwise or counterclockwise as desired, at varying speed, depending on the command signal. You might use a servo of this type on a radar dish if you mounted one on a robot. Or you could use one as a drive motor on a mobile robot.
- Linear servo: This is also like the positional rotation servo motor described above, but with additional gears (usually a rack and pinion mechanism) to change the output from circular to back-and-forth. These servos are not easy to find, but you can sometimes find them at hobby stores where they are used as actuators in larger model airplanes.

Selecting a Servo Motor

When starting a project that uses servos, look at your application requirements. How fast must the servo rotate from one position to another? How hard will it have to push or pull? Do I need a positional rotation, continuous rotation, or linear servo? How much overshoot is allowable? The less you pay for the servo, the less mechanical power it will have to muster and the less precision it will have in its movements. You can pay a bit more and get one that moves quickly, but it may not have a lot of power. You can also buy one that will pull or push large loads, but it may not move quickly or precisely. Manufacturers' websites and online hobby guides will have a lot of this information you can use to compare models. You will also find that hobby stores have a selection of servos and can usually help you decide which one is right for your project and budget.

Controlling a Servo Motor

Servos take commands from a series of pulses sent from the computer or radio. A pulse is a transition from low voltage to high voltage which stays high for a short time, and then returns to low. In battery devices such as servos, "low" is considered to be ground or 0 volts and "high" is the battery voltage. Servos tend to work in a range of 4.5 to 6 volts, so they are extremely hobbyist computer-friendly. Have you ever picked up one end of a rope that was tied to a tree or held one end of a jump rope while a friend held the other? Imagine that, while holding your end of the rope, you moved your arm up and down. The rope would make a big hump that would travel from your end to the other. What you have done is applied apulse, and it traveled down the rope as a wave. As you raise your hand up and down, if you keep your hand in the air longer, someone watching this experiment from the side would see that the pulse in the rope would be longer or wider. If you bring your hand down sooner, the pulse is shorter or more narrow. This is thepulse width. If you keep your end going up and down, making a whole bunch of these pulses one after another, you have created a pulse train (see Figure 6 below). How often did you raise and lower your end? This is the frequency of your pulse train and is measured in pulses per second, or Hz (abbreviation of "hertz").

Note: The microprocessor in your computer uses pulses from special clock circuitry to get the job done. Have you heard of your computer speed referred to as something like 1.7 gigahertz (GHz)? This is a way of saying that the pulses are coming at 1.7 billion pulses per second, or 1,700,000,000 Hz. Imagine trying to move your rope that fast!

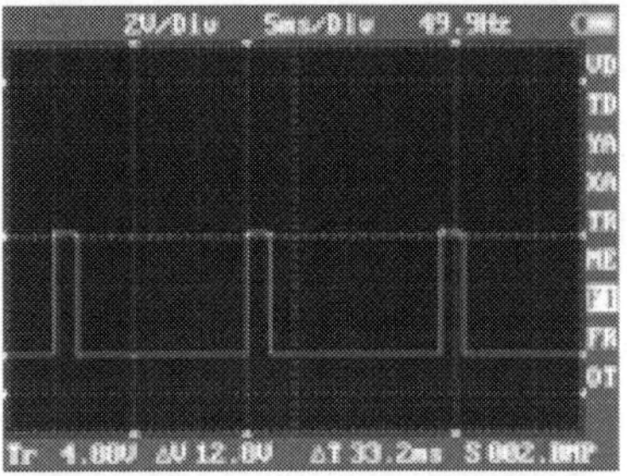

Fig. An example of a pulse train you might generate to control a servo, as shown in a screen capture from an inexpensive digital oscilloscope, an instrument for observing voltages). Here, a pulse is generated once every 20 milliseconds, or at about 50 Hz. In this example, the pulse width is about 2 milliseconds, which would have a servo rotate almost all the way to one end of its rotation. An oscilloscope is incredibly useful for testing and debugging systems that use servos.

Your servo must be connected to a source of power (4.5 to 6 volts) and the control signal must come from a computer or other circuitry. Each servo's requirements vary slightly, but a pulse train (as in Figure 6 above) of about 50 to 60 Hz works well for most models. The pulse width will vary from approximately 1 millisecond to 2 or 3 milliseconds (one millisecond is 1/1000 of a second).

Popular hobbyist computers such as the Arduino have software commands in the language for generating these pulse trains. But any microcontroller can be programmed to generate these waveforms. A system that passes information based on the width of pulses uses pulse width modulation (or PWM) and is a very common way of controlling motor speeds and LED brightness as well as servo motor position.

TYPICAL SERVO APPLICATION

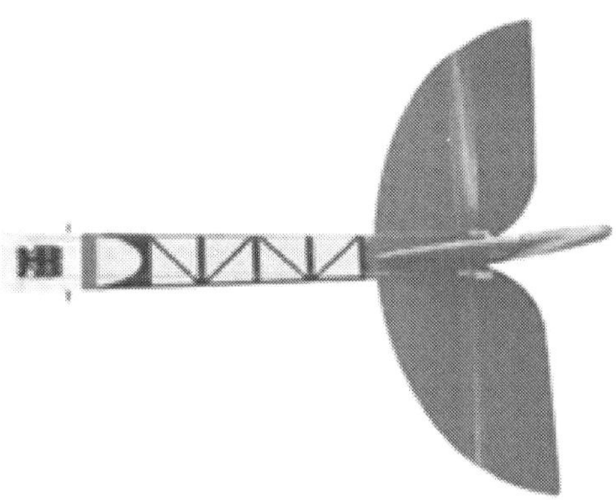

Fig. Two servos connected to rudder and elevator of a small RC plane.

Servos are typically used as part of a modular, radio-based, remote control system that provides one-way communication from an operator to the remote system, which might be a model helicopter or car. The basic goal is for the operator's manipulation of input controls on a radio transmitter to cause corresponding movements in the servos. Feedback about the success of the radio communication is generally limited to the operator's direct observation of the model, so these hobby radio control systems are typically used in open spaces and at distances limited by human sight.

Miniature cameras with wireless transmitters are available, but these are independent of the basic radio control system and are typically used more for the novelty of the remote point of view than for communication about the remote system's status. There are four basics elements to a standard radio control system:

- Transmitter
- Receiver
- Receiver battery
- Servos

The transmitter is on one end of a wireless link, and the receiver, its battery, and the servos are connected to each other in the device being controlled to make up the remote end of the wireless connection. (In cheap radio control toys, the three components on the remote end might not be modular and distinct, so there might not be a part to point to and call "the servo" or "the receiver".)

Although the non-servo components might be of limited interest to those not looking for a remote control solution, knowing a bit about the parts normally used with servos can provide some perspective about what to expect from servos, so we will cover the first three components in a bit more detail before moving on to servos.

Transmitter

Fig. Two-channel pistol grip RC transmitter.

The transmitter (sometimes simply called the radio) is the operator's user interface, typically with one or two joysticks and additional switches and knobs; car-specific "pistol grip" transmitters have a trigger for throttle and a small wheel for steering. Each degree of freedom on the model that the operator controls is operated by a different joystick axis, switch, or knob; the position of each of these control inputs is encoded on a separate channel. A minimal transmitter will use two channels, typically for throttle and steering on a car or boat.

Model aircraft often require four channels or more, with one joystick controlling throttle and rudder and another joystick controlling the elevator and ailerons; a switch on a fifth channel might control retractable landing gear, and a sixth channel might be controlled through a knob for variable deployment of flaps.

Each channel typically corresponds to a different servo to be controlled on the model. For instance, channel one and channel two might be on one joystick: moving it left or right would produce corresponding movements in the channel one servo (we'll get to what that means in a bit), and moving the joystick up and down would produce corresponding movements in the channel two servo.

The system being controlled might involve an arbitrarily complicated connection from the servo to the ultimate behavior being controlled, so it is important to have some means for calibration. For instance, we might know that a plane's rudder controls going left or right (yaw), but many factors affect how straight the plane flies, and accounting for all of them somehow without actually flying the plane is impractical. To address this need, the primary channels on transmitters usually have corresponding trim knobs or buttons to allow for slight changes to the correspondence between the control input's neutral position and the servo position. Thus, if a modeler on a maiden flight sees that the plane has a tendency to turn left, he can adjust the trim to the right rather than having to push the joystick to the right any time he wants to fly straight.

Fig. 6-channel RC transmitter.

A related concept is servo reversing, which allows the pilot to change the direction a servo moves for the same transmitter control input. In the normal state, the mechanical setup might happen to make the plane go left when the control stick is moved right; "reversing" the channel, typically through a small switch on the transmitter, will correct the problem.

Since the transmitter serves as the operator's user interface, it can have any number of extra features to allow for more advanced control, such as how far a servo travels for a given deflection of the control input or allowing one control input to control multiple channels. The point to keep in mind is that all of these features just change how the transmitter interprets the user's inputs, and changes nothing as far as the receiver or servos are concerned. One way to help understand this is to consider the case when you might need servo reversing as described in the previous example: your plane goes right when your control input goes left.

If you didn't have servo reversing, you could imagine removing the control stick from the transmitter, turning it 180 degrees, and putting it back in: left and right will now work properly, and nothing else in the transmitter or plane had to change. There is a lot more to RC transmitters, such as frequencies and modulations schemes, but those are not particularly relevant to using servos by themselves or to understanding the context of their intended application.

Receiver

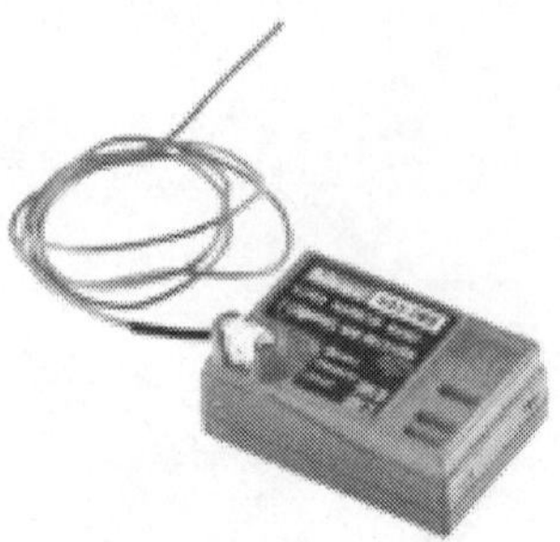

Fig. 2-channel RC receiver.

The receiver is the central component in the remote end of the RC system: it receives the radio signals from the transmitter, decodes them, and distributes corresponding commands to servos. Power for the servos is also routed through the receiver.

Receivers must be matched to their transmitters to allow for multiple systems to work simultaneously without interference; the details of the radio communication from the transmitter to the receiver are beyond the scope of this discussion. The important thing to note is that when we use servos as general-purpose actuators outside of the standard RC framework, we must substitute something for the receiver to tell our servos what to do.

Fig. 6-channel RC receiver.

Like transmitters, receivers are made for a particular number of channels, and receivers are equipped with a corresponding number of ports for connecting servos. The number of channels on a transmitter and receiver do not necessarily need to match, but the number of channels that can actually be controlled will be the lower of the two channel counts. A six-channel transmitter will periodically send out commands with positions for all six channels; the receiver picks up those commands and sends the channel one position command to the channel one servo port, the channel two position command to the channel two servo port, and so on.

The receiver also has a connector for applying power, which is used by the receiver and also supplied to the servos. For small receivers where size is an important consideration, the power connection is often combined with the connection for one of the servos.

Receiver Battery

The receiver and servos are powered by the receiver battery. Most of the power from the battery goes to the servos, so the battery must be sized to match the load the servos draw and to supply that current for a reasonable time. Low-end RC systems are sometimes supplied with a battery holder for four AAA or AA cells; using alkaline cells leads to a nominal voltage of 6 V, and using NiMH or NiCd rechargeable cells provides a nominal voltage of 4.8 V. The other common option is a battery pack of four or five NiMH or NiCd cells, which provides a nominal voltage of 4.8 V or 6.0 V, respectively. A higher voltage allows servos to provide more torque and speed, but some servos (typically very small ones) cannot withstand the higher voltage.

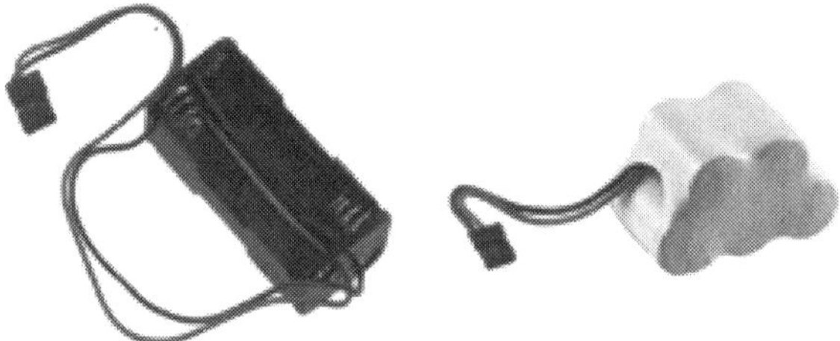

Fig. Battery holder for four AAA cells for RC receiver. 5-cell NiMH RC receiver battery pack.

Higher-capacity batteries tend to be larger and heavier than lower-capacity batteries, so the there is a tradeoff of remote system performance (especially in model aircraft) and battery life. However, the servos are usually used intermittently and do not need to strain much, so it is generally easy to achieve a battery life of several hours, which is usually much longer than the operator would want to operate the model and longer than most fuel supplies last. In the case of electric-powered models, a much larger battery is used for the main motor, whether it drives a propeller or drive wheels, and there are systems (often called "battery eliminator circuit", or BEC) for using this primary battery for the receiver and servos.

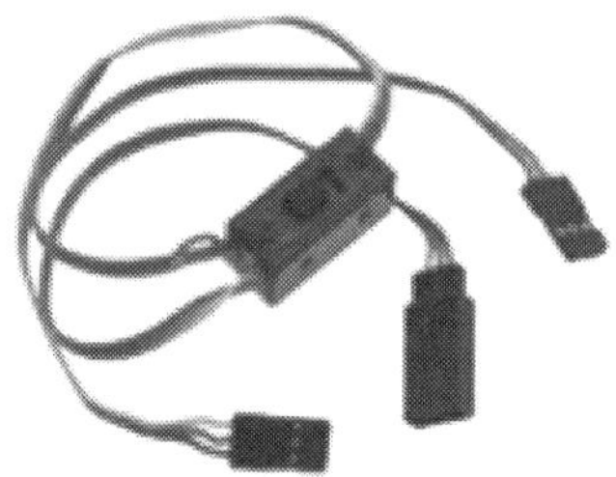

Fig. 12" (30 cm) servo switch harness.

Somewhat related to the receiver battery is the receiver switch harness, which is used between the battery and receiver to allow the remote system to be turned off. For some weight- or space-constrained applications, the switch might be omitted, and plugging in and unplugging the battery would be the only means of turning the system on and off.

Because radio control systems are designed to have the remote end unconstrained by power connections, there is no standard alternative to using a battery, whether it is a battery just for the radio system or a battery powering the whole model. Batteries are good power supplies for large and varying loads like the ones servos present, and finding economical alternatives to battery power can be difficult.

Servos

Fig. Standard-size servo.

Servos are the actuators in the standard RC system: they take the command from the receiver, use the electrical power from the battery, and physically move. The vast majority of servos have rotary outputs that move through a bit less than 180 degrees, though there are some specialty servos with linear outputs (the output moves straight back and forth, rather than rotating) and outputs that can turn a full rotation or more. What makes a servo a servo is that it takes care of getting the output to the commanded position by monitoring the output and applying power accordingly: if you push against a servo, it will push back to try to maintain its position. We will explore the details of what is inside a servo later.

Servos tend to be black plastic boxes with similar proportions, with the output shaft protruding on only one end, a 6-inch to 12-inch cable for connecting to the receiver protruding out the other, and mounting tabs along the sides. Some high-performance servos have metal cases for improved heat dissipation, and some servos have transparent cases. A "standard" servo has dimensions similar to those indicated in the following diagram:

This particular diagram is for a high-torque servo with metal gears, so its weight and output torque are higher than that of most standard servos, which might weigh around 1.5 ounces and provide around 40 ounce-inches of torque at a speed of about 0.2 seconds per sixty degrees. Standard servos are commonly used in 1/10 scale cars and model aircraft that weigh a few to a dozen pounds. Larger, "giant scale" or "1/4 scale" servos are available for larger cars and aircraft that might weigh several dozen pounds. Micro servos were initially around half the size and weight of standard servos, but improvements in electronics and manufacturing have led to even smaller, "sub-micro" servos that might weight less than a tenth of a standard servo. These servos are typically used in very light aircraft intended for indoor flight.

The following picture shows all components of the remote side of an RC system connected together. Giant, standard, and sub-micro servos are included to show their relative sizes, though it would not be common for these three sizes to be used together in a real installation.

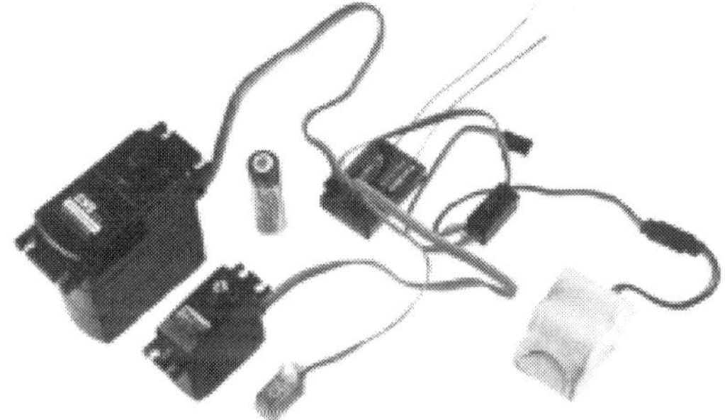

Fig. RC remote setup showing giant, standard, and sub-micro servos, receiver, receiver battery, switch, and AA cell for size reference.

Servo connectors have become much more standardized lately, so most servos are interchangeable (unlike transmitters and receivers, which use various frquencies and modulation schemes that are not interoperable). The vast majority of servos use 3-pin connectors with 0.1"-spaced pins and with identical pin assignments; the most common difference is in the polarizing method of the plug (designed to prevent plugging in the servo backward) and the wire color scheme.

Common Servo Specifications

The primary performance specifications for servos are torque, speed, weight, and size:

- Torque – In many applications, torque is one of the most important considerations. Generally, larger servos offer more torque, but expensive servos that use exotic materials can offer more torque without giving up any other features. Within a given size and price, some trade-off between torque and speed is usually available. Some manufacturers might just use completely different item numbers for different performance points; others might use a base model number with suffixes like "T" or "XF" to indicate the "high-torque" and "extra-fast" versions of otherwise identical servos. Standard servos typically provide forty to a few hundred ounce-inches; giant servos can get past 500 ounce-inches, and the smallest servos provide less than ten ounce-inches.

- Speed – Servo speed is usually specified in what might seem to be an odd manner: the time required to move sixty degrees. More generally common units for speed are rotations per second or rotations per minute (RPM), but those are not so useful for a device that does not make full rotations and spends a lot of its time accelerating. (If using a time for "speed" makes you uncomfortable, think of 100-meter sprinters: we report their performance in seconds, not miles per hour.) A typical servo speed is around 0.15 seconds per 60 degrees; less than 0.1 seconds is quite fast, and more than 0.25 seconds is noticeably slow.
- Weight – Weight is strongly correlated to output power (which is the product of speed and torque); going to higher speeds or higher torques (without corresponding drops in the other) will generally mean more weight. Standard servos weigh 1.5–2 ounces, giant servos can weigh five ounces or more, and micro servos weigh half an ounce or less. Small servos tend to use shorter and thinner wires to save weight.
- Size – Most servos have the same general shape, and the size generally scales with the output power and weight. However, there are some servos with noticeably different aspect ratios. These tend to be designed for specialty applications such as retractable landing gear or fitting in thin wings.

There are other specifications that cover aspects like materials used in the gears and the technology of the internal motor or electronics, but I'll leave those for another post. There are also many other parameters that robot builders might want to know but which are not specified; at least some of these missing specifications are because of the servo's intended use.

Ramifications of the Servo's Intended Use

While servos are great for their intended use, they have shortcomings as general-purpose actuators. It can be frustrating that there are thousands of different servos that all have the same limitations, so before wasting time looking for a servo that doesn't exist or denouncing a hobby store for not carrying it, it's good to consider why some things that matter to a robot builder might not matter to the RC hobbyist.

- Lack of feedback – Probably the most disappointing shortcoming of regular servos is that they provide no feedback. Since the main feature of a servo is its ability to get the output to the commanded position, it must know the output position, which could be very good information for other parts of a robot. Roboticists might want their servos to report all kind of other things, like how hard the servo is exerting itself to maintain a position or if there has been some kind of fault in the servo. Unfortunately, the basic paradigm of one-way communication from the radio control transmitter to the receiver means that even if a servo could somehow tell the receiver something, the receiver would still not be able to communicate that back to the remote operator, so there is no incentive for normal servo manufacturers to address this limitation.

- Twitch on power-up – A subtle problem related to the lack of feedback is that there must be some first command that a servo receives after being turned on. Since there are no commands for limiting speed or power, if the output shaft is not already in that commanded position, the servo will instantly try with all its might to head for the position. This doesn't really matter if it means a twitch in an airplane flap when it is turned on, but it can be quite dangerous for a robot arm or leg made of servos since trying to jerk a more massive mechanism could put a lot more stress on the servos. To limit the effects of this problem, it can be helpful to have a power-down position that the mechanism is put in before cutting power; on power-up, the first command can be to go to that position, and ideally, the mechanism will still be in that position and thereby minimize any self-destructive twitch. (Subsequent commands can move the servo slowly by sending position commands that are very close to each other, but there is no way of guaranteeing that the first command is for a position close to where the servo already is.)
- Limited rotation range – The standard rotation range being too limited is probably the second most common lament I have heard. With the exception of a few specialty applications like model sailboats, most radio control model applications just don't need more than about ±60 degrees of range. Most servos have about 150 to 200 degrees of range, so from the perspective of the manufacturers, that is more than enough for the necessary 120 degrees, and the exact limit on the range is usually not even specified. That can be quite annoying for a person who really wants a guaranteed 180 degrees.
- Low operating voltage – A relatively low operating voltage is a great feature for a system intended to use a few servos wirelessly for an hour at a time or less. It's less convenient for a hexapod walking robot or a continuously operating but stationary animatronic puppet since the low voltage means more current is

needed for the same amount of power. If you want to use twenty servos at once, you need a 20-amp power supply, which is not necessarily cheap, and you need fairly thick wires to handle the current efficiently.

- Short servo cable length – A requirement that servos be within a few feet of each other is not much of a limitation for a model whose largest dimension is not much more than that. It's tempting to use servos in larger-scale projects, but the low operating voltage and the command protocol used is not appropriate for just using longer wires. If you have a project in which you want servos spaced every ten feet or more, you will likely need to have some extra electronics at each servo so that the distance from the servo to the device it is plugged into is not more than a few feet.
- Current not specified – Since the currents used by servos are typically low enough, they are generally not specified. The problem is that a "low enough" stall current can add up if you have twenty servos straining at once, and a "low enough" idle current might not be low enough if you want to make a device that is powered all day but only needs to move at dawn and dusk. Part of the problem of the currents not being specified is that even if somebody helpful at Pololu measures one servo for you, you can't know how representative that is of the particular servo you will receive. (This is not an indictment of servo quality: for instance, all units of one servo model might be under a manufacturer's internal specification of 25 mA idle current; if we test one and it happens to be 5 mA and your servo is 15 mA, they are both good and close enough to zero for the intended use even though your unit is three times worse.) If you really care about your servo currents, you have to measure them yourself.

Fig. Servo connected to a mechanical speed control.

- Output shaft on only one side of servo – Normal servos made for the RC market have a short output shaft protruding on only one side of the case. The shaft is connected to an arm that usually pushes or pulls on some linkages, and having the shaft protrude from the other side would make the servo larger and more difficult to manufacture without much benefit for most of the intended customers. The one-sided output is not ideal for many robotics applications, and some specialty manufacturers offer brackets to enable creation of complex joints.
- Not designed for backdriving – Airplane modelers aren't really clamoring for a feature whereby they can pose their airplanes' control surfaces and have them "learn" the positions. (This is also related to the above mentioned limitation of no feedback.) Robot designers, on the other hand, might want to externally manipulate their creations. Unfortunately, most servos are not made for that, and some servos, especially the sub-micro ones with itty bitty gears, can reliably break if you try to back drive them (even when power is not applied).
- Audible noise – This might matter more to magic trick and kinetic sculpture creators than to most hobby robot builders, whose robot was never realistically going to be able to sneak up on the cat, anyway. Unfortunately for those looking for quiet servos, radio control folks using servos in models with loud engines and operating hundreds of feet away don't really care about how loud a servo is.
- Limited response or update rate – Servos are designed for real-time remote operation by humans, for which the standard update rate of fifty times per second is more than adequate. However, that might seem kind of slow for an autonomous robot which might be capable of updating its calculation of the desired servo position several hundred times per second. There are some servos, usually intended for use in conjunction with a gyro or other sensor on model aircraft, that can handle faster command rates.
- Limited options for larger servos – Servo manufacturers keep introducing newer servos, but the advancements tend to be in making micro servos smaller or in improving performance in standard servos. The same is not true for giant servos. Existing giant servos are already big enough for models with twelve foot wingspans and costing thousands of dollars, so there probably is not much of a market for even bigger servos. Also, the low-voltage limitation becomes a bigger problem the bigger a servo is and the more power it needs, so really big servos need some alternative and less standardized way of being powered.

SPECIALTY ROBOTICS SERVOS

Fig. Servo developed specifically for robotics applications.

Traditional servo manufacturers and manufacturers focusing on robotics have addressed most of the limitations listed above by developing specialized servos that have custom interfaces that allow for feedback and more advanced commands. Unfortunately, the extensions to the basic RC servo tend to be proprietary, so there is less to say about them in a general sense. Therefore, although those servos are made for robots and this series of articles is about using servos for robots, I will focus on traditional, standardized hobby servos. Most of the general information I provide should be relevant to robot servos, too, but keep in mind that if I mention some limitation of servos, there's probably some specialized robot servo out there that tries to address the limitation.

MODIFIED AND UNMODIFIED SERVOS

As it turns out there are two ways to use a servo. Most servos are used in applications where they are used to move to a specific position as in powering remote controlled vehicles. These are referred to as "unmodified servos". Unmodified servos, referred to in the WAM text, will go to a specific location when sent a pulse of specific width. Modified servos, included with the kit prepared by Parallax for the module, will be modified to allow for continuous rotation rather than moving to a specific location.

BOE-BOT NAVIGATION UNDER PROGRAM CONTROL

The Boe-Bot where to go and how to get there. You'll write programs to make the Boe-Bot perform a variety of maneuvers. Some programs can be used for navigating tight spaces, others for drawing shapes. Whatever the maneuver, this chapter presents the tools for programming the Boe-Bot to perform it. Here's what you'll learn how to do:

- Build a low battery indicator.
- Program your Boe-Bot to go a variety of directions, all in the same program.

- Write programs that fine tune the Boe-Bot's maneuvering skills.
- Write programs that remember long lists of movement instructions.
- Write programs that make the Boe-Bot accelerate and decelerate during maneuvers.

The variety of PBASIC programming techniques, such as for...next loops and if...then statements. The exercises in this chapter also offer lots of practice in using variables and flow control to accomplish a variety of tasks. Some essential math for converting program commands into distance and speed are also introduced. For some, this will be a first glimpse into elementary Dynamics.

CONVERTING INSTRUCTIONS TO MOTION

In the previous chapter, you programmed the Boe-Bot to move forward, backward, and turn in place. Additionally, software calibration settings were determined for programming the Boe-Bot to move straight forward, straight backward, and to stop and stay still. The Boe-Bot was programmed to go forward, it had to be reprogrammed to go backward and reprogrammed again to turn in place. In this chapter, all the directions will be incorporated into a single program. By determining how many pulses it takes to make the Boe-Bot rotate a certain amount during a turn, you can program the Boe-Bot to perform a variety of more precise maneuvers. For example, the Boe-Bot can be programmed to draw a square, or a cross, or a triangle.

This level of programmed maneuverability is well and good, but programming long and involved lists can become a complicated problem. PBASIC also features a simple and efficient method of recording and accessing long lists of directions in the program memory. You'll notice that while the Boe-Bot is performing its programmed maneuvers that it comes to abrupt stops when it changes direction. Commands can also be added to make the Boe-Bot decelerate into and accelerate out of direction changes. This will solve the abrupt stops and extend the life of the Boe-Bot's servos. The Boe-Bot's behavior when the batteries go low can be mystifying. Low batteries are also not good for the servos or the BASIC Stamp. The first activity will guide you through the construction and testing of a low battery indicator.

Low Battery Indicator

When the voltage supply drops below the level a device needs to function properly, it's called brownout. The BASIC Stamp protects itself from brownouts by making its processor and program memory chips go dormant until the power supply voltage returns to normal levels. A drop below 5.2 V at Vin results in a drop below 4.3 V at the BASIC Stamp's internal voltage regulator output. A circuit called a brownout detector on the BASIC Stamp waits for this condition. When the voltage comes back, the BASIC Stamp has been "reset." In response to a reset, the BASIC Stamp behaves essentially the same as if you had unplugged the power and then plugged it back in. In either case, what happens is that the program the BASIC Stamp was running before the reset starts over again at its beginning.

One way to indicate resets is to include an unmistakable signal at the beginning of all the Boe-Bot's programs. The signal then occurs every time the power gets plugged in, but it also occurs every time a reset due to brownout conditions occurs. The most effective signal for this is a speaker that emits a tone each time the BASIC Stamp program runs from the beginning or resets. The millisecond and microsecond quantities were introduced. In this chapter, the hertz and kilohertz quantities are introduced. One hertz is simply one time-per-second, and it's abbreviated 1 Hz. One kilohertz is one-thousand-times-per-second, and it's abbreviated 1 kHz.

Programming the Low Battery Indicator At the beginning of the program, a command that makes the speaker sound a tone will send the low battery signal. Then, the program listing should stay busy with an infinite loop until a reset occurs. If the circuit is correctly wired and the program is working, the tone should sound every time the power is unplugged then plugged back in. It should also sound every time the reset button is pressed. This guarantees that the tone will sound when a reset occurs due to low batteries.

- Connect the battery pack's barrel plug into the BOE's barrel jack.
- Connect the computer's serial cable to the BOE's DB9 connector.
- Run it by clicking Run then selecting Run.
- Verify that the low battery indicator works by pressing and releasing the reset button on the BOE.

Just as when the program first ran, the speaker should play a high pitched tone for 2 s. At the same time, the Debug Terminal should display the "Beep!!!" message. Then, all should go silent while the Debug Terminal displays line after line of the "Waiting for reset…" message.

- If the speaker did not play a tone, check your wiring and code, then try running the program again.
  Debug display problems are typically caused by typing mistakes.
- If the Debug display does not behave as expected, check your program and make sure the code matches.

Then, immediately after printing the message, the freqout command plays a 3 kHz tone on the piezoelectric speaker for 2 s. Playing the tone actually takes two steps. First, P2 must be set to output using the output 2 command. The audible tone is then played when the freqout 2, 2000, 3000 command is executed. It sends pulses out P2 that make the piezoelectric speaker vibrate at 3 kHz for 2 s. When the tone is done, the program enters an infinite loop, displaying the same "Waiting for reset…" message over and over again. Each time the reset button on the BOE is pressed or the power is disconnected and reconnected, the Program starts over again. The lines of code in the battery indicator program that generates the tone will be used at the beginning of every example program from here onward. You could consider it part of the "initialization routine" or "boot routine" for every Boe-Bot program.

CONTROLLING DISTANCE

Up to now, Boe-Bot programs have featured infinite loops. The Boe-Bot just kept going forward. This activity introduces a technique for controlling the distance the Boe-Bot travels.

Programming for Distance Control

An infinite loop can do a job, but it doesn't know when to stop. The best way to fix the problem is to replace the infinite loop with another kind of loop called a for…next loop. You can use a for...next loop to specify how many times the commands inside the loop are executed. A for…next loop uses a variable to keep track of how many times the commands inside it are executed. A for...next loop makes use of a variable to store a number that gets changed each time through the loop. In PBASIC, a variable has to be declared before it can be used.

An example of a variable declaration and an example of a for…next loop. The for…next loop is used instead of the infinite loop. The for...next loop makes the Boe-Bot go forward and then stop by controlling the number of pulses sent to the servos.

- Connect the battery pack's barrel plug into the BOE's barrel jack.
- Connect the computer's serial cable to the BOE's DB9 connector.
- Run it by clicking Run then selecting Run in the Stamp Editor.
- When the Boe-Bot speaker starts to beep indicating that the program is starting, press and hold the Reset button on the BOE.
- Unplug the serial cable from Boe-Bot.
- Place the Boe-Bot in the area you want it to navigate.
- Release the Reset button.
- Observe the Boe-Bot's behavior.

A variable named pulse_count is declared to be a word's worth of variable storage space using the command pulse_count var word. A word variable can store numbers between 0 and 65535. The handy thing about a variable is that there are a variety of PBASIC commands that can be used to change the variable's value. Other PBASIC commands can use a variable's value to make decisions.

Variables can also be used instead of numbers as arguments to certain PBASIC commands. You'll see examples of all these uses of variables in this chapter. The other options for variable declarations and the number ranges they can store. The size of the number a variable can store depends on how many bits it contains. A bit is a single, binary memory location that can either store a "1" or a "0." The more bits the larger the binary number you can store.

Table: Variable Declaration Sizes		
Size Declaration	Number of Bits	Can Store Numbers Ranging from-to
bit	1	0 to 1
nib	4	0 to 15
byte	8	0 to 255
word	16	0 to 65535 (or –32768 to +32767)

The four commands in the initialization routine should all be pretty familiar by now. The first two are the output and freqout commands used to signal when the program starts running. The initialization routine also includes the low commands that are used to set the initial output values of the I/O lines used to control the servos.

*Remember:*If the tone plays for no apparent reason when the Boe-Bot is in the middle of a maneuver, it indicates a brownout condition caused by low batteries. The main routine uses a for...next loop followed by a stop command. The for...next loop replaces the infinite goto loop used in previous examples. The commands nested in the for...next loop should be familiar by now. They are the commands that send the pulses to the Boe-Bot servos. The values of the period arguments used in the pulsout commands are the values for making the Boe-Bot roll forward.

The first time through the for next loop, the value of pulse_count is set to "1." Then the two pulsout commands and one pause command are executed. When the program gets to the next statement, it jumps back up to the for statement.

The second time through the loop, the for statement adds one to the value of pulse_count. Now the value of the pulse_count variable is "2," and the for statement checks to see if pulse_count is equal to 100 yet. Since pulse_count is not yet 100, the program continues to the next line. The three commands to pulse the servos and pause for 20 ms are executed again, and the next statement sends program control back to the for statement again. The value of pulse_count is incremented again and compared to the upper limit, and so on.

The 100th time the program gets to the next statement in the for...next loop, it sends the program up to the for statement again. This time, the value of pulse_count is incremented to 101. Now, when pulse_count is compared to the value of the end argument, it is greater than the upper limit of 100. So instead of continuing to the commands that send another pulse, the for statement sends program control to the command immediately after the next statement.

The stop command is executed immediately after the for...next loop, and it makes the program stop. The only thing that can get the BASIC Stamp out of a stop command is a hardware reset. In other words, if you want to run the program again, press the Reset (Rst) button on the BOE.

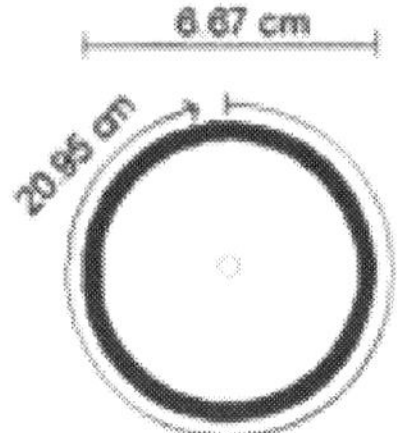

Fig. Wheel Diameter and Circumference

When programming the Boe-Bot, the goal is often to make it move a specific distance or to execute a particular turn. It is helpful to know how to figure out how far the Boe-Bot will travel or turn when it is given a specific command. Circumference is equal to pi (π) multiplied by the wheel diameter:

circumference = π x wheel diameter

circumference = 3.14159 x 6.67 cm ? 21 cm

So, now we know that with one complete turn of the wheels, the Boe-Bot travels about 21 cm. If we send pulses to the servo for the correct amount of time, the Boe-Bot can be made to travel a specific distance. This is because we can measure the speed that the Boe-Bot wheels turn. Once the speed is known, speed multiplied by time gives you distance. For example, with a pulsout command of 1000 delivered every 20 ms, the servo will turn at about 37.5 revolutions per minute (RPM), or 0.625 revolutions/sec. So the speed will be about:

21 cm/revolution x 0.625 revolutions/sec = 13.125 cm/s

The time it takes to make the Boe-Bot travel 50 cm is:

t_{travel} = 50 cm ? 13.125 cm/s ? about 3.81 seconds

Since the pulse and pause periods are known, the time it takes for a single loop can be calculated:

t_{loop} = 1.0 ms + 2.0 ms + 20 ms = 23 ms

Calculating the number of loops (the end argument for the for...next loop) is simply a matter of dividing tloop into ttravel.

Number of loops = 3.81 sec ? 23 ms/loop

= 3.81 s ? 0.023 s/loop ? 166 loops

- Modify Program Listing 2.2 for 166 forward loops (50 cm), then run it.

How far did the Boe-Bot go? Several factors can affect how far it moves, including differences between servos and battery voltage. However, the estimate of 166 is a good initial value to try. The for...next loop's end argument can then be fine tuned.

MANEUVERS – MAKING TURNS

If the same value is added to the center pulse width of one servo and subtracted from the center pulse width of the other, the Boe-Bot will travel in a straight line, either forward or backward. When the right servo gets a pulsout period of 500 (1.0 ms) and the left servo gets a pulsout period of 1000 (2.0 ms), the Boe-Bot goes forward. When the pulse periods for each servo are swapped, the Boe-Bot goes backward. If both servos receive 1.0 ms pulses, they turn in the same direction and cause the Boe-Bot to rotate counterclockwise. If many pulses are applied, the Boe-Bot will keep rotating. If 35 or so pulses are applied, the net effect is in the neighborhood of a 90? left turn. The same principles apply if both servos receive 2.0 ms pulses, except that the Boe-Bot will rotate clockwise instead of counterclockwise.

How the Turning in Place Program Works Only three changes were made to Program Listing 2.2 in order. First, the forward: and backward: labels were changed to left_turn: and right_turn: respectively. Then, the left_turn routine had both pulsout period arguments set to 500. The pulsout arguments in the right_turn routine were both set to 1000.

MANEUVERS – RAMPING

Ramping is a way to gradually increase the speed of the servos instead of suddenly making them go the opposite direction. This technique can increase the life expectancies both of your Boe-Bot's batteries and servos.

Programming for Ramping

The key to ramping is to use variables along with constants for servo pulse period. A for…next loop increments a variable each time the code nested between the for and next statements are executed. Since the value of this variable increases incrementally, it can be used to gradually increase the pulse widths.

How the Ramping Program Works

The pulses in the ramp_up_forward routine are nested in a for…next loop. Note the step 2 argument added to the for statement. None of the previous for...next loop examples used this argument, so the value of pulse_count was incremented in steps of 1. The step 2 argument in the for statement makes the value of pulse_count increment in steps of 2.

```
ramp_up_forward:
for pulse_count = 0 to 250 step 2
pulsout 12, 750 - pulse_count
pulsout 13, 750 + pulse_count
pause 20
next
```

The first time though the for...next loop, the value of pulse_count is "0." The second time through, pulse_count is incremented to "2," and the third time through it's "4," and so on, up to 250. The key to ramping is to modify the pulse period a little each time a pulse is sent to the servo until it's up to the desired period. Incorporating the value of pulse_count into the period argument of each pulsout command can be used to incrementally increase the pulse period in this way. Let's take a look at the pulses sent to the right servo. The command for sending the pulse is pulsout 12, 750 – pulse_count. Each time through the for...next loop, the value of pulse_count increases by two. This means that each time through the loop, the value of the pulsout command's period argument is decreased by two, since pulse_count is subtracted from 750. By the last pass through the loop, the value of pulse_count is up to 250, which is subtracted from 750 to give a value of 500. For the left servo, the value of pulse_count is added to 750, and the last time through the loop gives a period of 1000. When these target values are reached, the servos are up to full speed forward, and the program can move on to the familiar forward routine.

When the forward routine is finished, the challenge is to ramp the pulse widths back down to the resting value of 750. A useful feature of PBASIC is that for...next loops count down instead of up if the start argument is larger than the end argument. This method was used for counting back down in the ramp_down_forward routine. The for...next loop started pulse_count at a value of 250, then counted down to 0 in steps of 2. This is because the command for pulse_count = 100 to 0 step 2 was used.

REMEMBERING LONG LISTS USING EEPROM

The BASIC Stamp stores its tokenized version of the PBASIC program in electrically erasable programmable read only memory (EEPROM). Physically, the EEPROM is the small black chip on the BASIC Stamp II module labeled "24LC16B." This part is made by Microchip Inc. The BASIC Stamp's EEPROM can hold 2048 bytes (2 kB) of information. What's not used for program storage (which builds from address 2047 toward address 0) can be used for data storage (which builds from address 0 toward address 2047).

FYIIf the data collides with your program, the PBASIC program won't execute properly. EEPROM memory is different from RAM variable storage in several aspects:

- EEPROM takes more time to store a value, sometimes up to several milliseconds.
- EEPROM can accept a finite number of write cycles, around 10 million writes. RAM has unlimited read/write capabilities.
- The primary function of the EEPROM is to store programs; data is stored in leftover space.

You can view the contents of the BASIC Stamp's EEPROM by clicking Run and selecting Memory Map.

This program might have seemed large while you were typing it in, but it only takes up 128 of the available 2048 bytes of available program memory. There currently is enough room for quite a long list of instructions. One of the simpler approaches is to store characters indicating which way to go. Since a character occupies a byte in memory, there is currently room for 1920 one-character direction instructions.

SIMPLIFY NAVIGATION WITH SUBROUTINES

Programming Navigation with Subroutines

A subroutine is a segment of code that does a particular job. To make the subroutine do its job, a command is used in the main routine that "calls" the subroutine. The command for calling a subroutine is the gosub command, and it's similar to the goto command. A goto command tells the program to go to a label, and then start executing instructions. The gosub command tells the program to go to a label, and start executing instructions, but come back when finished. A subroutine is finished when the return command is encountered.

How the Subroutine Navigation Program Works

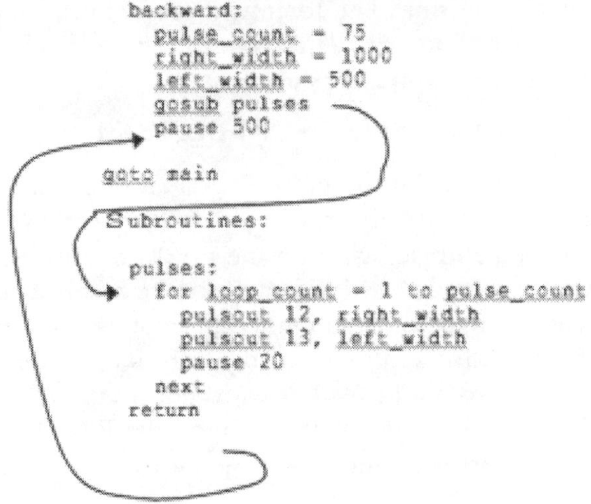

Fig. Program flow for subroutines.

The backward routine calls the pulses subroutine. The commands in the pulses subroutine are executed until the program gets to the return command. The return command sends the program back to the command just after gosub pulses command, which is pause 500 in this case. This same process occurred in the forward routine, which was executed before the backward routine shown.

ALL TOGETHER NOW

This chapter has introduced a variety of navigation techniques:

- Navigation routines that control distance and turning
- Speed ramping

- Remembering long lists
- Calling subroutines
- Using variables to set values used by subroutines

All these elements now can be tied together to make a more robust navigational program.

HOW EEPROM NAVIGATION WITH RAMPING WORKS

Six word-size variables are added to the declarations. The variables old_left and old_right are for remembering the previous pulse period for each servo. The variables right_gap and left_gap are variables that change slightly each time a pulse is applied. When the gap is large, the pulse width is closer to the old pulse values. When the gap is small, the pulses are closer to the new pulse periods.

The data directive has been restructured so that it stores a direction, followed by a number of pulses. This way, the Boe-Bot can be given directions like: go forward 125 pulses, then backward 125 pulses, then right 75 pulses, and so on. Initial values are set for the variables that keep track of pulse widths. They all start with the center value of 750. The first command in the main routine is gosub pulses. Since the initial number of pulses to deliver is a value of "0," and the initial values of the old and current servo instructions are at center, the pulses routine will do nothing the first time through. Every time through the main routine after that, the gosub command calls the pulses subroutine, which pulses the servos with values determined from the previous pass through the main routine.

There are now two read commands in the main routine instead of just one. This is so that both the direction and number of pulses can be read. The first read command still accesses a letter that denotes a Boe-Bot direction. The second read command looks up EE_address + 1, which is the location of the number that follows the direction in the data statement. This is a handy way to access the pulse count value in a single command. Then EE_address is incremented by two instead of one. This way, the next time the first read command looks for a letter, it will skip over the number (of pulses) and find the next letter in the sequence.

Now that a letter indicating direction has been stored in the instruction variable and the number of pulses have been stored in the pulse_count variable, it's time for the BASIC Stamp to figure out what that letter stored in instruction means. Instead of using a series of if...then statements, two commands are used. The first is the lookdown command. The lookdown command will take the value "B" from instruction and find that it's the first in the list. Actually, as far as the lookdown command is concerned, it's really the zero entry in the list since the lookup command starts counting from zero. So, it stores the number 0 in the direction variable. If instruction is an "L," the number 1 will be stored in direction. If instruction is "R," the number 2 will be stored in direction, and so on. The branch command skips to a label based on the variable's value. If direction is "0," branch will skip to the backward routine. If direction is "1," branch will skip to the left_turn label. If direction is "2," branch will skip to the right_turn label, and so on.

You may not have been aware that you could place more than one command separated by colons on a line. The backward, left_turn, right_turn, forward, and quit routines each occupy a single line. The only problem that can occur when multiple PBASIC commands are on a single line is that a label cannot be found in the middle of a line, only at the beginning. Each of the direction routines sets the value of right_width and left_width period values, just as in previous programs. Instead of calling the pulses routine, each of the four direction routines ends with a goto main command. Note that as soon as the program gets to main, it calls the pulses subroutine.

The pulses subroutine first determines the difference between the previous value sent to it for each servo and the current value. Taking the right servo as an example, the variable that stores this value is right_gap, because it stores the current right_width minus the old_right width. The process is the same for the left servo, except the sign is reversed. Next, the for...next loop applying pulse_count pulses is started.

Assuming that right_width is different from old_right, the right_gap value will start large. Since right_gap is subtracted from right_width when the pulsout command is executed, it stands to reason that the right_gap value should initially be large. Then, with each pulse, the right_gap value should get smaller and smaller as the pulse width ramps up to its target value. A single command takes care of decrementing right_gap if it's positive and incrementing it if it's negative. Right_gap = right_gap - 5.

How does this expression work? If right_gap is a positive number, right_gap.bit15 will be zero. So, if right_gap is positive, it still gets 5 subtracted from it. However, if right_gap is negative, right_gap.bit15 is "1," not "0." This is because of how the two's complement binary system works in the BASIC Stamp. This means that right_gap also gets 10 added to it for a net gain of 5. The same principle applies for the left_gap.

Next the pulses are delivered. Note that each period argument is an expression. For the pulsout command to P12, the period is right_width – right_gap, and for the period delivered to P13, it's left_width – left_gap. It's important to keep track of what very last pulse values delivered to the servos were before fetching the next navigation instruction from EEPROM. The two commands after the for...next loop that delivers pulses save the pulse widths.

old_right = right_width - right_gap

old_left = left_width - left_gap

The next time the program returns to the pulses subroutine, the first two commands set the initial values:

pulses:
right_gap = right_width-old_right
left_gap = left_width - old_left

The for...next loop works in the pulses subroutine gradually adjusts the pulse values from where they started to the target value.

REAL WORLD EXAMPLE

When the Boe-Bot's batteries go below a certain level, the Boe-Bot sends a signal indicating that something went wrong by sending electronic pulses to a piezoelectric speaker. These speakers are common. Think of the number of appliances that beep when you press a key or do something. Your microwave oven, grocery store cash registers, and alarm system keypads all have this feature.

It should come as no surprise that each has a microcontroller monitoring the keys on a keypad. Level sensors are also common in industry, where voltage, pressure, temperature and a variety of other conditions have to be controlled. Microcontrollers are often used to monitor these conditions and make an alarm speaker sound when the process goes outside acceptable levels.

Micro-controlled motion is also all around us. Although there may not be that many autonomous rolling robots in your household yet, there are many other gizmos with micro-controlled moving parts. Printer heads and computer disk drives are two examples that use stepper motors. Servos controlled by microcontrollers are also used in a variety of places.

Many automobile systems rely on servos to control small moving parts in various engine and emission systems. Industrial servos maintain many factory processes, often in conjunction with the level sensors discussed earlier.

BOE-BOT APPLICATIONS

Programmed navigation is the foundation for a variety of other Boe-Bot activities. In the Projects section, you'll work on programming the Boe-Bot to navigate a variety of shapes and on fine tuning some of the navigational algorithms. Some of these routines to respond to sensor inputs.

PROJECTS

One aspect of Boe-Bot navigation that was not covered in this chapter is drawing curves. A curve occurs any time the left wheel is not turning the same speed as the right wheel. Curves also occur when the left wheel is accelerating/decelerating at a different rate than the right wheel. Keep in mind that the number of pulses delivered is a way of controlling distance for a given pulse width (speed).

- The Boe-Bot is being used to transport reactive material, in particular solid sodium and water. If the two react, they explode, leaving your Boe-Bot as a pile of components. In order to carefully transport the chemicals, you will need to start movements with gradually increasing velocity. Create a program that drives the Boe-Bot in a one-meter square with smooth turning transitions.

- Create a simple movement pattern with several directions, for example F,B,R,R,F,F,L, and lastly F. These patterns would be stored and read from the EEPROM. When you are done executing the pattern, trace the same pattern backwards.
- Create source code for the following movement patterns:

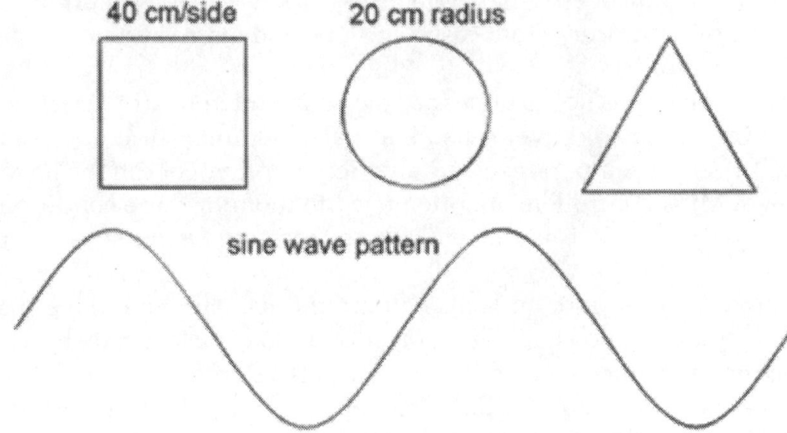

Fig. Boe-Bot paths to program.

TACTILE NAVIGATION

The Whiskers kit is so named because the kit's bumper switches look like whiskers, though some argue they look more like antennae. At any rate, whiskers give the Boe-Bot the ability to sense the world around it with tactile inputs.

The Boe-Bot can use these whiskers to navigate only by touch. Although the activities in this chapter focus on using just the whiskers, they can also be used with other sensors to increase the Boe-Bot's functionality.

BUILDING AND TESTING THE WHISKERS

Parts

- 10 kΩ resistors
- 3-pin headers
- 3/8″ 4/40 male/female standoffs
- ¼″ 4/40 machine screws
- Boe-Bot bumper wires
- Nylon washers size #4

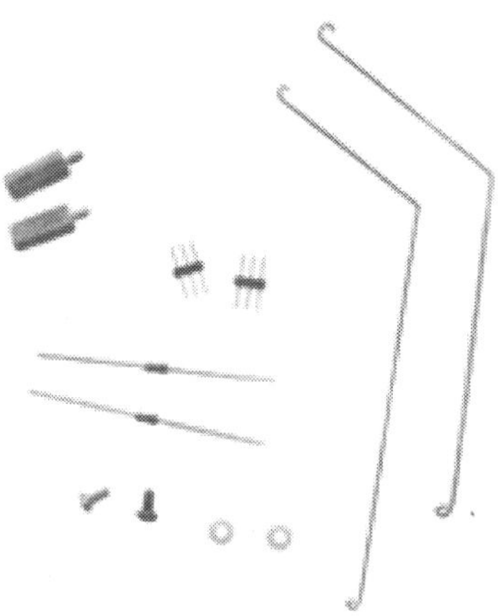

Fig. Whiskers parts

Build It! Your #1 point Phillips screwdriver and quarter-inch combination wrench will come in handy here. Before getting started on whisker construction, take a close look at Figure 3.2. Use these pictures as a guide while constructing the mechanical part of the Whiskers kit. Figure 3.3 shows the whiskers wiring diagram. Follow it for making the necessary electrical connections.

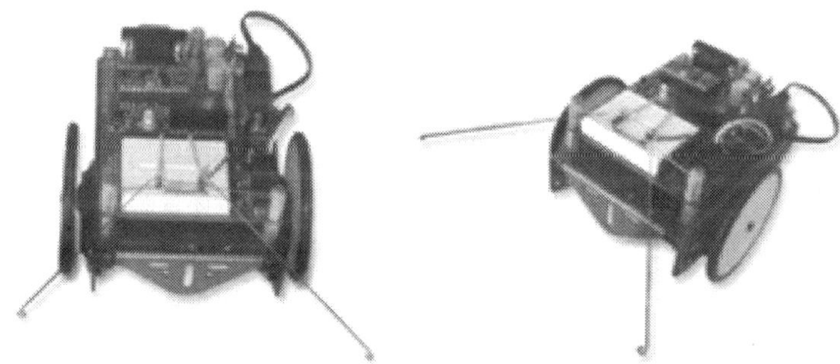

Fig. Pictures of Boe-Bot with Whiskers

- Remove the two front screws that hold your Board of Education to the front standoffs.
- Screw in the male/female standoffs included in the Whiskers kit in place of the screws that were just removed.
- TIP
  Hold the male/female standoff by the servo port while turning the standoff between the BOE and Boe-Bot chassis to tighten it. The standoff between the BOE and chassis won't turn until you loosen the screw that holds it to the chassis. Make sure to retighten it when you're done.
- Place a nylon washer on top of each standoff.

- Thread each screw removed in the first step through the open loop of a whisker.
- Screw each screw into a standoff sandwiching the loop of the whisker between the screw head and the nylon washer.

Whisker Inputs

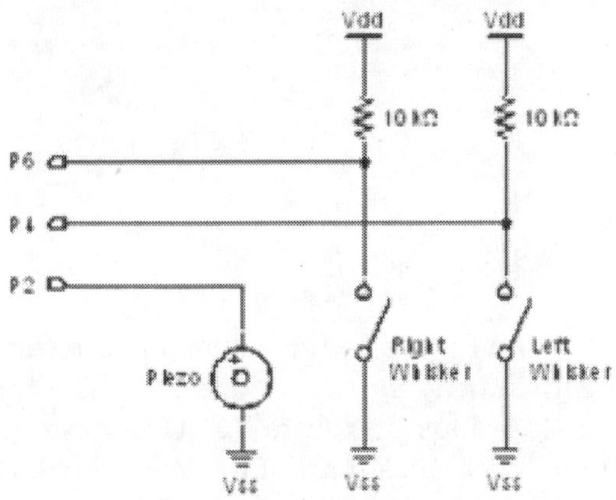

Fig. Whiskers Schematic.

A schematic representation of the circuit you've just built. Each whisker is both the mechanical extension and the ground electrical connection of a normally open, single-pole, single-throw switch. The BASIC Stamp can be programmed to detect when a whisker is pressed. I/O pins connected to each switch circuit monitor the voltage at the 10 kΩ pull-up resistor. When a given whisker is not pressed, the voltage at the I/O pin connected to that whisker is 5 V (logic 1). When a whisker is pressed, the I/O line is shorted to ground, so the I/O line sees 0 V (logic 0).

SERVO PROBLEMS WITH ARDUINO

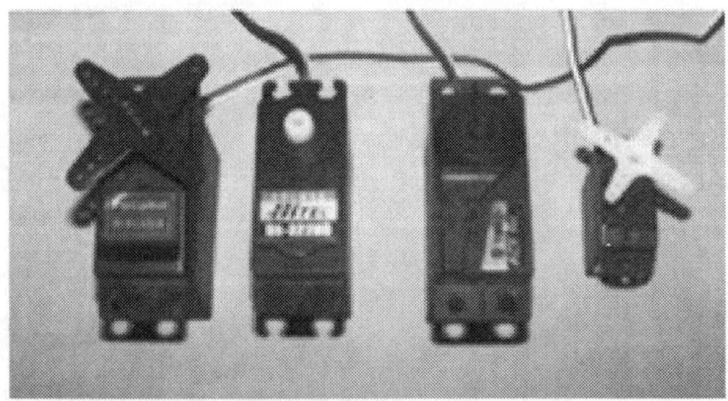

Servo problems are one of the most frequently posted topics within the Arduino community. While problems may arise from programming, circuit design and faulty hardware, the vast majority or problems are a result of insufficient power or incorrectly connected power. Control is nothing without power ! The Arduino itself is very good at controlling servos, the Servo library will allow a single Arduino to control upto 12 servos with no additional hardware.

What the Arduino cannot do is deliver power to 12 Servos, its questionable whether an Arduino can reliably deliver power to even a single servo. Power - Not All Servos Are Created Equal: While building the circuit for the pictures in this post, I decided to measure the current drawn by each of the test servos with no load applied. Each of the servos is the standard size used in radio controlled cars, they are all low end servos available for between 10 and 15 dollars. No Load Current Of Test Servos In mA

The Bluebird draws five times more current at no load than the Futaba, its also less smooth and the least expensive of the servos tested. The current at no load is only relevant in a small number of applications, all of the servos drew more than the the maximum 250mA I was able to measure when subjected to a light load in the form of finger pressure. I will be interested to follow up this test in a future post using a variety of repeatable loads to compare servo performance and current draw. The fact that the lowest cost servo draws five times more current than the Futaba servo suggest that this will be an important concern for larger autonomous projects which will need to operate both under their own power and under load. Note that while controlling all four servos the Arduino drew only 10mA from its separate power source.

So How Can Successfully Drive Lots of Servos With An Arduino ?

The power problem is easily solved through the addition of a 'power circuit'. This is can be as simple as four disposable AA Batteries such as you might use in a camera or toy car.

Fig. 4 Domestic AA Batteries in a holder

Rechargable AAs are an even better option, they store as much charge, deliver as much current and can be used again and again. For large projects which need to operate under their own power the basic concept is the same however the choice of battery technology will be different. See the 'Performance Power' section in Part 2. So why do we need two power circuits ? The Arduino has a narrow operating voltage around 5 Volts (3.3 Volts in some) and is sensitive to variation in this voltage. The Arduino design is based on the assumption that a stable 5 Volt power source will be feeding the chip at the heart of the Arduino. In the case of the popular UNO, this regulated 5 Volt power is supplied by the USB Connection or through a regulator built into the board.

The onboard regulator is designed to provide power to the Arduino and supporting circuitry. It is not designed to power external devices and trying to do so is the single most common reason for failure with servo projects. This 'Not designed to power external devices' also applies to USB Connected projects. The remainder of this post provides a walk through of setting up an Arduino servo example which will drive four servos from the Arduino using a separate power pack to meet the power requirements of the servos.

Arduino Servo Walkthrough -

Power The Arduino

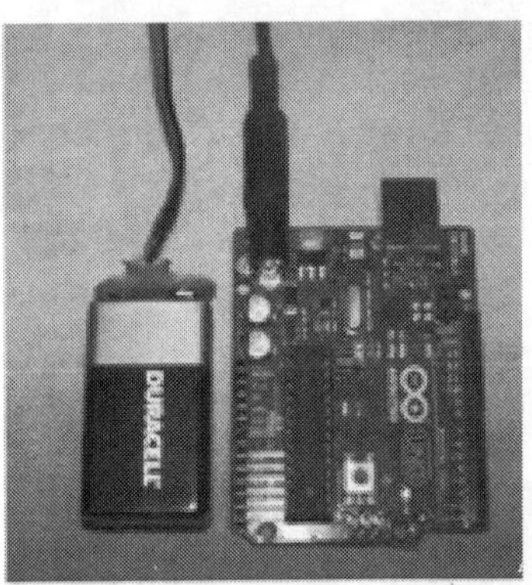

For the walkthrough I am using an Arduino UNO powered by a PP3 9 Volt Battery. The Arduino has been loaded with my Multi Sweep example sketch, a link is provided at the end. While running the servos, the Arduino was drawing only 10mA. A good PP3 could power the Arduino for days, after all, they power smoke alarms for months. See the note below for more on the PP3 and why its not a good battery for use elsewhere in your projects. Note - The PP3 is a poor battery choice for most applications, it has a small charge capacity (run time) and cannot deliver the higher currents required to drive servos or motors, however the 9 volts it provides is great for powering an Arduino through the onboard regulator. As the Arduino makes so little demand on a battery the PP3 is a common and practical choice to power the Arduino - just not any shields, motors, servos, transmitters etc.

The PP3's small charge capacity and limited ability to deliver current make it an unsuitable choice for providing the power circuit in our projects, the common AA battery is a far better alternative. In the case of servos the 9

Volts supplied by an unregulated PP3 is over the 4.8 to 6 volt recommended operating range and will result in immediate damage to the servo. Again AA Batteries are a better choice as four will provide a usable 6 Volts for our servo power circuit and a better run time. This wont work for me, I need USB for Serial Output - This is no problem at all, you can simply connect the USB Cable to the Arduino as you normally would. This will provide power to the Arduino so you do not need to use the 9V PP3 Battery.

You should still use the separate servo power and this will work provided that the ground from the battery pack is connected to the Arduino.

Power For The Servos

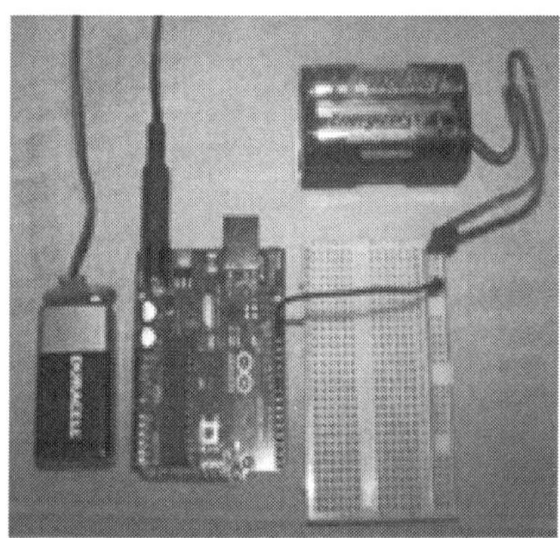

For servo power I am using four disposable AA Batteries. These are high capacity versions sold for cameras which will give us enough current and charge (run time) for our servos. Each AA Battery provides 1.5 volts for a total of 6 volts (4*1.5). Most servos are designed to operate with 4.8 to 6 volts. Powering them with more voltage can result in instant damage. The four AA Batteries give us a usable 6 volts. I have connected the AA Batteries to the power (red) and ground (black) rails on my bread board.

Notice the black jumper running from the bread board to the Arduino ? This is the next most common mistake in servo projects, when people introduce the servo power pack, they forget to connect a common ground. You must connect the ground wire between the power circuit and the Arduino (control circuit) without this connection your circuit will not work. This applies whether you are powering your Arduino from a wall socket, USB port or a battery.

In the picture you can see where I have simply connected the two circuits through the black jumper wire running from the black ground rail of the bread board to the ground (GND) pin on the Arduino next to pin 13. All of the ground pins on the Arduino are connected so use which GND pin is most convenient. These two circuits now share a common ground allowing us to add some servos.

Connecting Individual Servo Power

I have added a 3-Pin section of PCB Header to make it easier to connect the female servo plug to female breadboard. Next to this you can see that I have added a green jumper from the battery pack ground rail to pin1 of the header, this is the shared ground between the Arduino, battery pack and servo. Next I have added a yellow jumper from the power rail - 6 Volts from the AA Battery pack - to the center pin of the header. This will provide power to our servo directly from the AA Battery pack, not from the Arduino or its 9 Volt PP3 Battery.

The final pin is the signal pin, this is the pin which the Arduino will use to tell the servo which position to hold. Remember - Without the common ground between the Arduino and the Servo AA battery pack, your project will not work. In the picture, this common ground is provided by the black wire linking the ground rail of the breadboard with the GND pin next to pin 13 of the Arduino.

Adding a servo

Fig. All we need to do now is connect our first servo -

You can see here that I have connected the first servo taking care to make sure that the black ground wire of the servo connector is connected to pin 1 of my header. This is then connected to the shared ground of the Arduino (and Battery pack) through the green jumper wire. You can also see that I have added headers for three more servos following the same convention of yellow for 6Volt power from the AA Battery pack and green for the shared Arduino/AA Battery pack ground. The servo is receiving its position signals from the Arduino through the white jumper wire connected to pin 13. Remember - Without the common ground (black wire) connecting the Arduino GND Pin to the ground (-) fo the battery pack through the black ground rail of the breadboard, this will not work ! If you have uploaded the multi sweep example sketch linked at the end of this post, you should see your servo sweeping back and forth.

Adding more servos

Once you have one servo up and running, you can add more by following the same convention - connect the servo plug to the header so that the end with the black wire attaches to the header pin with the green wire.

4

Satellite and VSAT: Innovative Uses for Rural Telephony and Internet Development

Nearly all of Africa's international bandwidth is provided by satellite. Except for those countries which are connected to and utilising submarine fibre-optic cables (Algeria, Djibouti, Egypt, Morocco, Senegal, South Africa, Tunisia, Canary Islands and Cape Verde), satellite presents the only means of carrying international other than terrestrial or microwave links they might have with neighbours. As a result, African countries have a very high dependency on satellite, with the majority of countries more than 95% of international traffic carried by satellite.

SATELLITE BASICS

SATELLITE NAMES AND NUMBERS

Television communication satellites are located in a stationary orbit above the earth. All satellites are designated by a name and number. Lined up across the sky in what is commonly known as the domestic satellite arc, these satellites are designated by their positions in the sky, which are given a "degrees west" designation. For example, Telstar 401 is located at 97 degrees west. A visual display of the domestic satellite arc is shown below and in Appendix A. These satellites are owned and operated by corporations that sell or lease time on different channels to broadcast users. For example, the satellites in the TelStar series are all owned by AT&T. Other satellites are in the SpaceNet series, the Galaxy series, and the SatCom series, owned by different companies.

NTC buys transmission time on different satellites. Many times the availability of transponder time dictates which satellite NTC will use for its broadcasts. Other agencies and commercial vendors may transmit telecasts on any satellite throughout the satellite arc. To easily receive NTC programs and broadcasts from other programmers, field offices with downlink satellite equipment should have a good working knowledge of satellite transmissions and reception. Make sure you have the correct satellite name, number, and location when positioning your dish for reception.

Channels and Transponders

Every satellite has many transponders. Most transponders can carry two programs at once by manipulating their polarization; this is referred to as having two channels. Therefore, a satellite with 12 transponders will actually have 24 channels. The words used to designate the correct transponder are "horizontal" or "upper" and "vertical" or "lower." The resulting wording of broadcast announcements will be similar to these examples:

Transponder 10/horizontal, Channel 19 - which means 10/horizontal is channel 19 Transponder 18/vertical, Channel 22 - which means18/vertical is channel 22

- Or -

Transponder 12/upper, Channel 15 - which means 12/upper is channel 15

Transponder 17/lower, Channel 18 - which means17/lower is channel 18

You need to understand the difference between a transponder and a channel. Remember, receivers tune by channel, so that is the most important information. It's a plus to know which transponder is involved because of differences in receiver tuning. But concentrate on knowing the channel. Your satellite receiver should also be programmed correctly so that you can easily receive all channels on all transponders.

Frequency

Frequency is the radio frequency the signal travels on. You won't normally use this information, but some receivers show the frequency, and you can use this display to double check that you're tuned to the right channel. Both video and audio signals have frequency designations.

Satellite Bands

The two satellite frequency bands, C-Band and Ku-Band, are comparable to AM and FM on your radio. Just as you can't hear an FM radio station on an AM-only radio, you can't receive a Ku-band program on a C-band dish. BLM offices have bought satellite dishes that can receive both C-Band Ku-Band transmissions. Sometimes programs are broadcast simultaneously on both bands so that any downlink can receive them. Receiving C-band is usually easier, as the dish aiming is less critical. But C-band is susceptible to local electrical or microwave interference. Ku-band is less susceptible to electrical interference, but extremely heavy rains over the receive site can briefly disrupt Ku reception.

Most home backyard satellite dishes are C-band only. Many educational or industrial sites have downlinks that receive both C and Ku. If you are using an offsite facility for your downlink, check to be sure they can receive the correct satellite band for the broadcast you will be viewing, especially if it is only on Ku-band. Do not accept someone's vague impression of whether the downlink is compatible with your broadcast. Confirm everything with those who know what they are doing. If no one knows, go there yourself, tune in the satellite you will use, and look at any active channel on the satellite to witness the downlink's performance.

Another reason to check the site's capability is that individual downlinks may have problems seeing a particular satellite, especially satellites with sky positions near the local horizon. Buildings, mountains, or trees may also block the view of a satellite from a specific downlink. Your office should subscribe to a publication such as Satellite TV Week or the Keystone Satellite Guide. These publications give up-to-date information on satellite locations, frequencies, transponder positions, and launches of new satellites.

Keystone Communications' web site also contains current satellite location information (http://www.keystonecom.com). Even though satellites are in service for many years, programming and positioning regularly change. For example, NTC is purchasing some of its broadcast time on Satellite Galaxy 9 or G-9. This is a newer satellite that was launched in May 1996 into position 123 degrees west, in the position previously occupied by Telstar 303. Telstar 303 or T-3 has been moved to a new location. Many older receivers have not been reprogrammed to reflect these changes.

Scrambled Signals

Some satellite broadcasts are scrambled or encrypted so that only authorized users can view them. Normally, NTC does not scramble its broadcasts, but offices may wish to view commercial telecasts that are scrambled. To receive scrambled signals from commercial telecasts, offices need a compatible decoder on their satellite receivers. Viewing these programs requires the following steps:

- Confirm that the receiver at the viewing site has a decoder, and find out what the decoder ID number is.
- Call the program producer and give the downlink decoder's ID number. During the test signal broadcast before the program begins, the producers will feed a digital code specific to the ID number of each authorized receiver's decoder. This code enables the decoder to receive the program.
- If your decoder does not bring in the program clearly during the test period, call the producer's telephone help number and ask for your decoder to be re-enabled. The digital code will be fed again.

Time Zones

Time zones can be confusing, and it is not always obvious what time the transmission will be broadcast in your time zone. NTC will always notate local Phoenix and Eastern time on its broadcast schedules. Most commercial vendors publish their broadcast schedules using Eastern time. Always convert the broadcast information to your local time. Remember, Arizona does not switch to daylight savings time, so consider this fact in your calculations. During the summer NTC is on the same time as the West Coast Pacific Daylight Time (PDT). Technically this is also Mountain Standard Time, but it is not the same time as Mountain Daylight Time.

BEFORE THE BROADCAST

Once you decide to subscribe to a satellite program, you must make an up-front investment of time and effort. Both technical and logistical preparations must be considered. This up-front work will help eliminate last-minute problems and ensure a successful session for participants.

Determining Audience Size, Event Announcements, and Publicity

NTC will send announcement memos to field offices. These memos will include course description and objective(s), dates, times, and nomination due dates for all broadcasts. Local offices should help spread the word to employees about upcoming broadcasts. Posted announcements, e-mail messages, and discussions at staff meetings are ways that managers and local coordinators can help ensure that people who may benefit from training and informational broadcasts are aware of the date and time of each event. Critical to the success of any teleconference are announcements that give the course description & objective(s), date, and time of the teleconference. Send announcements to your anticipated viewing audience as far in advance of the program date as possible. If the event is an NTC training session, employees must be nominated for the course through the NTC training nomination process. If the BLM event is not a formal NTC training course, you will receive a program announcement that can be posted or sent to employees via GroupWise. When announcing a program, request that participants arrive at least 15 minutes prior to the start of the teleconference to receive participation information. Once the number of participants in each broadcast at your site has been determined, coordinators should proceed with all on-site logistics.

Communication From NTC

For most telecasts NTC will send field offices written materials for participants and facilitators before the broadcast. This material will include the following: The Broadcast Schedule, which contains all broadcast times, the program agenda, scheduled breaks, planned call-in sessions, and test signal information. Technical Information, which includes the satellite name, position, downlink frequency, and the transponder and channel number.

Longer events and multi-day training courses may be broadcast on different satellites or transponders. Appendix B contains a sample NTC broadcast announcement.

Help Hot Line Phone Numbers, which are the telephone numbers downlink sites should call during the transmission of test signals if they are having problems with receiving NTC's signal. Site Facilitator Information, which will include instructions on how local facilitators should work with groups at field offices. Instructions on Interactivity, which will give information on the use of phone calls and faxes related to each broadcast. A duplication master of the NTC teleconference fax form is provided in this handbook. Downlink Site Guide, to help facilitators and others who are coordinating local events before, during, and after the broadcast. Student Notebooks. In many cases each office will receive one set of master documents. On-site coordinators will need to copy these materials for each participant at field offices. Course-Specific Materials, which may include maps, computer discs, mineral samples, reference books, or other publications.

Participant Material Preparation and Distribution

Some training courses that use satellite broadcasts will also have pre-assignments. Copy and distribute these materials to all participants in a timely manner before the broadcast to allow time for the work to be done. More information on pre-assignments will be sent out from NTC with the prebroadcast materials. Assemble all packets of materials for participants in advance and have them ready for trainees as they arrive or register. The packets might contain the following:

- Name tags
- An agenda
- A program schedule
- Textbooks or workbooks
- Program-related handouts
- A pencil and pad of paper
- The program evaluation form
- The NTC teleconference fax form
- A list of participating sites
- Telephone instructions for call-in questions
- List of instructors (including local presenters) with biographies

Signal Reception and Technical Arrangements

Unless you are making the technical arrangements yourself, give your support staff copies of the technical satellite broadcast information, including the Satellite Help Hot Line phone number. Each receive site should check to see that all satellite equipment is working, that someone from the office knows how to use it, and that these people will be ready on the day of the broadcast.

More than one person should be familiar with the equipment in an office in case the primary operator is ill or called away unexpectedly on the day of the broadcast. These staff members should also be able to handle any last-minute satellite tuning, video recording, trouble shooting with the audio systems, and adjusting video monitors or projectors.

A day or two before the broadcast, pre-focus your satellite downlink to the satellite NTC will be using, and test your signal reception from that satellite. To ensure that you are on the correct satellite, check your reception against published satellite programming guides such as Satellite TV Week. These guides give information on other programming that is broadcast on each satellite and transponder.

Audiovisual Equipment Arrangements

Depending on the size of your local group, arrange for video monitors, audio amplification equipment, or video projection equipment. For training telecasts ensure that everyone has a clear view of the screen and that graphics and other training visuals can be easily seen. Check prebroadcast materials sent out from NTC to see if more audiovisual equipment is required for use during the training. Such equipment may include slide and overhead projectors and VHS video players. Be sure that you have spare lamps and any other technical support you may need.

Recording Broadcasts

If you plan to video tape the broadcast at your office, have a ½-inch VHS video recorder wired into your system. Make a test tape of a signal feed off of your satellite receiving equipment before all NTC broadcasts. You may want to video tape the broadcast for later use or for people who cannot attend the live telecast. Make sure you have enough blank VHS video tapes ready to record broadcast segments. Prelabel tapes so they don't get mixed up during taping.

Before recording any satellite broadcast, make sure you have the legal right to do so. If the telecast is free, you can probably video tape it with no legal problems. But ask first! If you have to pay to receive the program, videotaping will probably be restricted. You may be required to ask for permission to tape the program and pay an additional fee. As the registered client, you are legally responsible for not violating copyright laws. If BLM originates the broadcast, you can safely assume that you can tape it and use it for additional training. You can videotape almost all programs produced by federal agencies. If in doubt call the program producer for clarification.

Telephone Arrangements

If the teleconference will include call-in interactive segments, the room must be equipped with telephone lines over which long-distance calls can be placed by FTS, credit card, or toll-free 1-800 numbers.

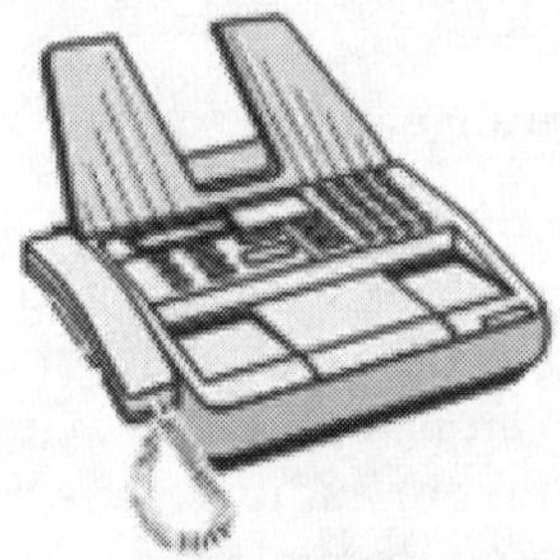

To allow participants to call in questions and comments during the broadcast, arrange to have a telephone in your viewing room. Place the phone as far away from the audio source of the conference as possible, ideally on the other side of the room from the monitor or projection screen. To reduce echo and feedback, locate this phone away from all speakers, and do not use a speaker phone. To avoid these problems, some sites have used cellular phones out in the hall directly next to the viewing room.

To permit participants to fax in questions and comments during the broadcast, arrange to have a fax machine in or close to your viewing room. Be sure to have paper in the machine in case your site receives incoming faxes during the broadcast. If possible, use a dedicated phone line for your fax communications. For special high-profile events, NTC also sends the audio portion of the program to a telephone bridge. Such a bridge gives receive sites access to the program audio should satellite downlink equipment fail. Receive sites that want to be ready to use this backup signal should have a speaker phone placed in their viewing room, in addition to the telephone that is used for placing calls to the NTC studios. Information and instructions for backup audio feeds of NTC teleconferences will be included in prebroadcast materials sent to receive sites before the event.

Cautions About Communication Security

Coordinators, facilitators, and subject matter specialists are responsible for maintaining the security of the NTC telephone and fax number for all broadcasts. In most cases these numbers are not distributed to the public or to anyone outside the target audience for each broadcast. To avoid unwanted calls and having the NTC phone and fax lines jammed, please do not release these phone numbers to anyone not authorized to participate in the interactive portion of the broadcast. In many cases other people may watch telecasts but do not call or fax in communications.

Choosing a Meeting Room and Arrangement

As soon as a broadcast is announced, schedule your receiving room for that day. Determine the arrangement of tables, chairs, and television viewing equipment in your room. The layout needed for a viewing site depends on the number of viewers and the activities involved. Set up the tables and chairs well before the event to meet the needs of the training session and your local participants. Depending on the number of participants and the surrounding activities, you may need to rent more tables and chairs.

If the broadcast involves hands-on computer training, you may need to install DOS or UNIX computer equipment in your viewing room. Decide if you will need signs or posters to direct participants to the building or classroom. Prepare these signs ahead of time so that they are ready to be posted on the day of the event. A small conference room will handle up to 30 people for a lecture-type program. But if the program involves group activity and meal breaks, you may need a larger room. The type of facility depends on the training requirements and the needs of facilitated group activities. If you add local activities to your sessions, you may need more space.

Sample Room Arrangements

An effective room arrangement can facilitate group interaction and learning. The purpose of the program should be considered when choosing the room arrangement. If the training session or program contains small group exercises, layouts one or two (shown on the following pages) are recommended. In these arrangements, the participants are already arranged in smaller groups so when the exercise is announced, they can concentrate on the task at hand rather than moving into small groups. If the task requires even smaller groups, each table can split into two groups.

INTRODUCTION TO SATELLITE ACCOUNTS

This lecture note is intended to introduce satellite accounts as complementary to the System of National Accounts (SNA). Topics and discussions address the questions like what is the basic difference in the approach of satellite accounts verses the SNA central framework?, What are Satellite Accounts?, What Satellite Accounts deal with and allow for?, What are Ancillary Activity Satellite Accounts?, What are Functionally Oriented Satellite Accounts?; and other important matters concerning Satellite Accounts?

SNA Central Framework vs Satellite Accounts

SNA central framework has an integrated accounting structure that is exhaustive and consistent within boundary of economic activities it covers. The set of concepts adopted by the system is fully coherent. The central framework may be used in a flexible way in order to put greater or lesser emphasis on specific aspect of economic life such as public sector, household, high inflation. However the margins of flexibility allowed by the central framework do not permit conflicting approaches to be covered simultaneously Counterpart of the benefits of central framework is that there are certain limitations as to what may be accommodated directly by it against what all a user may want. Satellite accounts or systems generally stress the need to expand the analytical capacity of national accounting for selected areas of social concern in a flexible manner without overburdening or disrupting the central system.

Typically satellite accounts allow for the:

- Provision of additional information on particular social concerns of a functional or cross sector nature.
- Use of complementary and alternative concepts including use of complementary and alternative classifications and accounting framework when needed to introduce additional dimensions to the national accounts.
- Extended coverage of costs and benefits of human activities.
- Further analysis of data by means of relevant indicators and aggregates.
- Linkage of physical data sources and analysis to the monetary accounting system.

Broadly two types of Satellite Accounts are popular, one which focuses on details of an aspect without expansion of SNA framework for example satellite accounts for Agriculture, Tourism, NPIs. Health, Education; and the other type making expansion of the framework beyond the boundaries (production or assets or both) of the SNA for example satellite accounts for Environment, Household, Human Resources, etc.

The Satellite Accounts Framework is derived from SNA as the central framework for specific type of analysis and may expand the framework beyond the boundaries (production or assets or both) of the SNA. The Satellite Accounts are designed to expand the analytical capacity of "basic" economic accounts without overburdening them with details or interfering with their general purpose orientation. The Satellite Accounts are meant to supplement rather than to replace the existing accounts, organize information in an internally consistent way that suits their particular analytical focus, while maintaining links to the existing accounts. The Satellite Accounts may typically expand a particular segment of the existing account with additional information, including non-monetary. They may use definitions and classifications that differ from those in the existing accounts. The Satellite Accounts may modify the production and/ or asset boundary of the SNA, depending on the analytical focus.

Satellite Accounts allow for provision of additional information on particular social concerns of a functional or cross sector nature. They may allow for use of complementary or alternative concepts, including the use of complementary and alternative classifications and accounting frameworks. Satellite Accounts may also allow for extended coverage of costs and benefits of human activities and further analysis of data by means of relevant indicators and aggregates. Certain satellite accounts allow for the linkage of physical data sources and analysis to the monetary accounting systems.

Satellite Analysis

Satellite accounts play a dual role as tools for analysis and statistical coordination. Satellite accounts deal with analysis of production and products, primary incomes and transfers, uses of goods and services, assets and liabilities, and purposes as well as aggregates. For illustration of production and products, consider the case when establishments, and consequently industries are not homogeneous at a given level of ISIC, they undertake both principal activity and secondary activities. The output of secondary activities is identified according to its nature but the inputs of secondary activities are not separated from the ones of principal activities. Ancillary activities on the other hand are not analyzed and classified according to their own nature. This means that ancillary activities are not undertaken by distinct producer units and the related products do not appear as autonomous products.

For illustration of primary incomes and transfers consider the case when production boundary is extended, primary income is increased as income being imputed for the additional activities which are inserted in the production boundary. Additional analysis is also possible for primary incomes, for example one may separate the mixed income of households into element of return to labour and an element of return to capital. Several kinds of transfers in addition to those in central framework may be delineated, if meaningful. For example the case of non-market service provided free of charge by government units to market producers. In the central framework these services are collective consumption of government. If a further analysis treats them as an addition to intermediate consumption of market producers, a counterpart should be introduced, preferably in subsidies on production.

The coverage of uses of goods and services, either for intermediate consumption or final consumption or capital formation, obviously changes as a result of enlarging the concept of production. The borderline between intermediate consumption, final consumption and capital formation may also be modified in various ways. For example when final consumption in education and health is treated as fixed capital formation; the corresponding central framework transaction is reclassified from consumption to capital formation which results in human capital assets. The scope of assets and liabilities is also modified as a consequence of extending the concept of production or modifying the borderline between consumption and capital formation as indicated above.

In the central framework an analysis by purpose is applied to most transactions of general government and NPISH to households. As many times purposes of current interest hardly feature in existing general classification, priority is given to the analysis of other purposes. For example, besides the fact that tourism does not appear as such as a main category in the classification of household goods and services by object, it is not even reasonably possible to reassemble the necessary pieces because not all of them are shown in the classification. The main aggregates may be modified due to various complementary and alternative analyses mentioned above.

Thus broadly speaking two types of satellite analysis may be distinguished in their relationship with the central framework of SNA- One type involves some rearrangement of central classifications and introduction to complementary elements that differ from the conceptual framework (such as the identification of the output of ancillary activities) without drastically diverging from the concepts on which the central framework is built. The second type of satellite analysis is mainly based on the concepts that are alternatives to the ones in the SNA. A different production boundary or enlarged concepts of consumption and capital formation may be introduced, or the scope of assets may be extended, the borderline between economic phenomena as covered by the central framework and natural phenomena may be altered, etc.

Ancillary Activity Satellite Accounts

Ancillary activities (central administrative office, warehouse, garages, repair shops, transport section, in house electric power plants) are not undertaken by distinct producer, they are undertaken for the internal use of the unit/ establishment/enterprise. As already discussed units conducting ancillary activities are not regarded as separate producers even if located separately in SNA. Ancillary activities therefore do not appear as autonomous products. In the Ancillary Activity Satellite Accounts therefore the ancillary activities are considered as autonomous products and shown first as production and then as consumption. This helps in identifying the contribution and role of the ancillary activity product in the economy.

Functionally Oriented Satellite Accounts

Framework for functionally oriented satellite accounts concentrate on one field to give full picture of it, in a systematic way, by establishing a specific accounting framework, articulated with the central framework. Satellite framework does not aim at covering all economic life; it is a self consistent framework in a partial domain. A satellite framework or account is by hypothesis more functional rather than institutional. To put more emphasis on the functional point of view, such satellite accounts combine an extension of the kind of activity and product analysis and a generalization of the purpose approach. Such accounts are relevant for many fields, such as culture, education, health, social protection, tourism, environmental protection, research and development (R&D), development aid, transportation, data processing, housing and communications.

Scope of a functionally oriented satellite account that consider a specific field in depth while preserving the possibility of calculating some significant aggregates such as national expenditure, the starting point is an analysis of the uses. This corresponds to the questions 'how many resources are devoted to (education, transportation, etc.)?' or, 'how much is spent (education, transportation, etc.)?'. In order to answer these questions, we have to decide upon (i) goods and services specific to the field, (ii) activities for which capital formation is required, and (iii) transfers specific to the field.

Depending on the field, the design of a given satellite account will emphasize on

- Detailed analysis of the production and uses of the specific goods and services
- Detailed analysis of transfers (e.g., Social Protection)
- Both production/uses and transfers equally (e.g., Education, Health)
- Uses as such (Tourism, Environmental Protection)

Matters concerning Satellite Accounts

Analyzing in detail, who are users, consumers, investors, or transfer recipients: is an important part of the satellite account. For example in national expenditure the users are units which actually acquire goods and services (actual final consumption, intermediate consumption or capital formation) or receive specific transfers which are not intended to finance these acquisitions of goods and services.Analysis of uses or benefits out of national expenditure: In satellite account in a given field covers the analysis of uses or benefits out of national expenditure, production and its factors, transfers and other ways of financing the uses both in value terms and when relevant, in physical quantities.

Components of Uses/ National Expenditure:

Components of Uses or national expenditure distinguish two types of specific goods and services, the characteristic goods and services and connected goods and services. The first category covers the products which are typical for the field under study. Interest is in what way these goods and services are produced?, what kind of producers are involved?, what kinds of labour and fixed capital they use? and the efficiency of the production process?, thus the allocation of resources. For example, for health, characteristic products are health services, public administration services, education and R&D services in health.

The second category, connected goods and services includes products in whose uses we are interested because they are covered by the concept of expenditure in the given field without being typical. For example in health, transportation of patients may be considered connected goods and services. Capital formation in specific goods and services: is a part of the components of uses/ national expenditure. In a satellite account for housing, for example, it covers fixed capital formation in residential building. In R & D, Education and Health this item will be empty if the related services are all included under consumption as in central framework and if all or part of these services were treated as capital formation in a satellite account, the corresponding uses would be under capital formation in specific goods and services In an account for culture, acquisition less disposal of valuable such as paintings may be significant. It is also a component of capital formation in specific goods and services.

National Expenditure by Components and by Users/ Beneficiaries

For users or beneficiaries the terminology used may differ from one satellite account to another. 'Users' is more relevant to tourism or housing whereas 'beneficiaries' to social protection or development aid. In both cases the terms refer to who is using the goods and services or benefiting from the transfers involved.

At most aggregated level the classification of users/ beneficiaries is simply a rearrangement of the central framework classification of institutional sectors and types of producers in which production aspect and consumption aspect are separated. It may be:

- Market producers
- Non-market producers
- Government as a collective consumer
- Households as consumers
- Rest of the World.

Households, even individuals as consumers are the most important type of users/ beneficiaries in many satellite accounts. In order to be useful for social analysis and policy, a further breakdown of households (size of income, age, sex, location, etc) is necessary.

National Expenditure by Components and by Financing Units

As users do not always bear the expenses themselves it may be necessary to analyze the financing units, i.e., the units which ultimately bear the expenses. For this purpose a classification derived from the central framework classification of institutional sectors, such as the following:

- Market production
- NPISHs
- General government
- Households
- Financial enterprises
- Rest of the World

May be used for a satellite account giving the cross classification of components of national expenditure by financing units.

Non-monetary Data

A satellite account allows for the linkage of physical data to monetary accounting system. In central framework this kind of link remains generally implicit, being done only for population and labour inputs where these figures are necessary for calculating a number of per person indicators such as income or consumption and indicators of productivity. Non-monetary data are more important for user/ beneficiaries of goods and services and recipients of transfers. These data are especially meaningful in the fields of social concern such as education, health and social protection. They are indispensable to asses the standard of living of various parts of the population and to look in depth at redistribution policy. Physical data are not to be considered as a secondary part of a satellite account. They are essential components, both for the information they provide directly and in order to make the monetary data fully meaningful.

Environmental and Economic Accounting

Environmental and Economic Accounting considers environmental analysis in the context of a broadened framework that amends several concepts of the SNA to respond to the growing concerns of incorporating environmental criteria in economic analysis. In these accounts, SNA aggregates are amended to treat natural resources as capital in the production of goods and services, to record the cost of using, i.e. depleting and degrading those resources and to record the implicit transfers needed to account for the imputed cost and capital items.

In SNA many costs and capital items of accounting for natural resources are identified separately in the classification and accounts dealing with stocks and other volume changes of assets. These features of SNA facilitate use of the system as a point of departure for environmental accounting. However several elements of the SNA particularly those for the account for the other volume changes need to be broken down further and reclassified and other elements have to be added for the specific purposes of environmental accounting.

In the SNA only produced assets including inventories are explicitly taken into account in the calculation of net value added (NVA). The cost of their use is reflected in intermediate consumption of fixed capital. Non-produced natural assets such as land, mineral resources and forests are included in the SNA asset boundary in so far as they are under the effective control of institutional units. However the cost of their use is not explicitly accounted for in production cost. This may either imply that the price of the products does not reflect such cost or, if it does, as may be the case for some depletion cost- such cost is not separately identified but lumped together with other unidentified elements in the residual derivation of operating surplus. Three main approaches to environmental accounting: first one natural resource accounting focusing on accounts in physical terms; second liked to national accounts is environmental accounts in monetary terms; and third is welfare oriented one. The second one, monetary satellite accounting identifies the actual expenditures on environmental protection and deals with the treatment of environmental cost to natural and other assets caused by production activities in the calculation of net product. Monetary satellite accounting is generally more limited in coverage of environmental concerns than physical resource accounting. The welfare oriented approach deals with the environmental effects borne by individuals and by producers other than the producers causing these effects.

The physical resource accounting is most advanced in terms of practical implementation. Monetary satellite accounting has some controversies with regard to valuation. The least consensus exists with regard to the welfare approach to environmental accounting.

Natural Resource Accounting in Physical Terms

Natural resource accounting focuses on physical asset balances i.e., opening and closing stocks and changes their in of materials, energy and natural resources. Where applicable (for selected pollutants) it may also include changes in environmental quality of natural assets in terms of environmental (quality) indices.

Environmental Accounts in Monetary Terms

Monetary environmental accounts in restricted sense identify within national accounts the actual expenditures on environmental protection. Monetary environmental accounts also include the functional approach to environmental accounting. Further in the monetary environmental accounts GDP is adjusted for selective environmental costs, including the cost of oil depletion, deforestation, depletion of fish stock, and the cost of soil erosion. The System of Environmental Economic Accounting (SEEA) does not distinguish between depletion and degradation but rather quantitative and qualitative use of natural assets. The SEEA considers effects on production analysis, identifying the environmental cost of depletion and degradation caused by different economic activities and showing the corresponding effects on natural and other assets.

Welfare and Similar Approaches

Instead of dealing with the cost caused by production activities and their effects on capital used in production, welfare approach focuses on the environmental impacts of cost borne or, in a broader sense, on the well being. The approach considers the free environmental services provided by nature to producers and consumers and the subsequent damage borne by them. The environmental services provided free and the damages borne are implicitly considered as transfers by and to nature, which increase or decrease environmentally adjusted net national income.

Another approach is based on concept of environmental sustainability standards and on estimating the necessary avoidance or restoration costs to meet these standards.

Asset Boundary and Classification

The most important amendment introduced into environmental accounting as compared to SNA is the extension of the asset boundary. In SNA, natural assets are included only if they provide economic benefits to the owner.

Thus only economic assets (those having institutional ownership) are considered. However, in environmental accounting both economic assets and environmental assets are considered in order to fully cover depletion, degradation and other accumulation. i.e., transfer of natural assets to economic activities. Classification of natural assets in SEEA differs to that of in SNA. SEEA asset classification covers wider range of natural assets and environmental impacts on them, including ground water depletion and air pollution. Of course, more selective environmental accounting could focus on only those natural assets that are economic assets.

Environmental cost

The SEEA identifies two types of environmental cost. The first is imputed cost for degradation and depletion. The second is the actual cost incurred in the form of environmental protection expenses

Use of Non-produced Natural Assets

The use of non-produced natural assets is introduced as an additional cost in the SEEA. They represent depletion and degradation in physical terms - e.g., quantity of minerals extracted, quantity of timber cut, volume of solid, liquid or gaseous wastes generated, or alternatively monetary allowance for depletion and degradation.

Environmental Protection Expenses

In addition to the disposal services provided by the environment free of charge in the case of degradation, actual expenses are incurred to avoid environmental degradation or eliminate the effects after degradation takes place. Increasingly enterprises explicitly produce such services on a commercial basis. In many instances however the services are produced as the ancillary activities. SEEA recommends classifying and compiling and accounting the environmental protection expenses in the SEEA as per CEPA (Classification of Environmental Protection Activities).

System of Economic Accounts for Food, Agriculture (SEAFA)

System of Economic Accounts for Food, Agriculture (SEAFA) recommended by Food and Agricultural organization (FAO) is designed for analysis of products from the agriculture sector. It is designed to focus on Food supply and demand and to link agriculture with other components of economy for planning and policy making. Key Aspects of SEAFA involve the scope- what products (Goods and services), Economic activities and transactions, Institutional sectors and transactions to be included while basically following the SNA Framework as a core.

Why SEAFA as a separate system? SEAFA is a separate system as a satellite account for the several reasons. Main reasons include: SEAFA intends to focus on the basic needs of human population for food, fiber, fuel and shelter supplied by a combination of agriculture, fishing and forestry activities; SEAFA main interest is in food production, impact of food and agricultural activities on eco-system/natural resources; SEAFA requires information on labor force, Research and Development, balance of trade, BOP, depletion of natural resources; Food production is not specifically identified in ISIC/ SNA; and for SEAFA Food availability is required in terms of nutritional levels.

Products and Economic Activities Included in SEAFA

Products included in the scope of SEAFA are Agriculture products, Forestry products, Fishery products, Manufactured food products, Environmental protection services, and Other related services. Economic Activities included in SEAFA are Agricultural crop and livestock production, Forestry, logging and related activities, Fishing, Manufacturing (ISIC Division 15), Environmental protection activities, Agriculture-related services: agricultural research, veterinary services, education, manufacturing of fertilizers, insecticides/pesticides, agriculture capital industries.

Tourism Satellite Account (TSA)

Tourism Satellite Account – Recommended Methodological Framework (TSARMF) is the guide for compiling the TSA. Salient features of TSA include the following: Tourism Satellite Account (TSA) identifies the economic aspects of tourism separately, but still within, the core NA framework. The tourism 'industry' is defined not by the suppliers of its output, but rather by the consumers of its output. As such, the tourism industry, as measured in a TSA, cuts across potentially all industries in the core national accounts. It is not readily apparent because "tourism" is not identified as a conventional industry or product. Tourism is however implicitly included in core national accounts statistics. TSA refers basically to demand of goods and services of Visitors from the country and from other countries. TSA involves interaction of the demand and supply with other activities and transactions in the economy. The aim of a TSA is to identify the value of tourism expenditure (Visitor Consumption) and the proportion of each industry's output (Tourism Industry Ratio) that is consumed by visitors/tourists.

SPACE SEGMENT

The major operators of satellites over Africa are: Anatolia (Kalitel), Europestar, Eutelsat, Intelsat, Lockheed Martin, New Skies, and PanAmSat. These operators' fleets of satellites have overlapping footprints in both C-band and Ku-band which between them cover every inch of the continent. However, the existing satellites above Africa are heavily subscribed. There is simply not much capacity left, and it is increasingly difficult to lease capacity on them. However, new satellites are being launched - notably Stellat 5 in May 2002, and New Skies Satellites will launch NSS 7 on 16 April. These satellite operators' current supply and future demand justify the huge expense of launching new satellites. The Intelsat launches (903 on 30 March, 601 in July) for example comprise part of US$3bn investment to expand the fleet by 10 satellites to 24. Where new satellites are not being launched, some operators are repositioning existing satellites in their fleet over the region to provide additional coverage.

Upcoming satellite launches:

- Stellat 5.
- New Skies Satellite NSS 7.
- Intelsat 601.
- Intelsat 801.
- Intelsat 903.

Behind the launch of these commercial satellites lies the distant promise of the Regional African Satellite Communications Organisation (RASCOM) project. Established in 1992 by an intergovernmental treaty now signed by 44 countries, RASCOM was originally formed to pool and optimise satellite capacity leased from Intelsat, to help maximise the discounts by buying transponder space in bulk. A second phase is to see the launch of a dedicated satellite for the region; a build-operate-transfer (BOT) partnership was formed to launch RASCOM Star, but although this was due to be operational by 2003 it has yet to get off the ground. The concept behind was to establish direct links between all African countries and to provide infrastructure on large scale to connect rural areas, with thousands of VSAT terminals.

South Africa is interested in launching its own regional satellite. This alternative was a key factor in Telkom's disengagement from investing in the RASCOM project, which needs the biggest carrier on the continent to help underwrite the cost. Instead, South Africa is believed to be looking at two alternative proposals: SACOMSAT (South Africa Communications Satellite) and SADC-1, a sub-regional project jointly owned by the 14 members of the Southern Africa Development Community (SADC) region. Nigeria, too, has ambitions to launch its own satellite and flirted with the idea of its own space programme.

EARTH SEGMENT

Until recently, the main customers for satellite-based connectivity were public telecommunication operators (PTOs), chiefly because they held monopolies and were the gatekeepers of international traffic. These built earth stations for international traffic, and in a few cases also to connect far-flung parts of the country to the PSTN. For example, Telkom South Africa has (in conjunction with Hughes Network Systems) deployed more than 34,000 VSATs across the country roll out the PSTN to rural areas and also corporate users. With the process of liberalisation underway, Africa is a mosaic of licensing and new telecommunication service providers are emerging. International voice traffic being is increasingly being carried by licensed satellite service providers and mobile operators, and data traffic by Internet service providers (ISPs) where they are licensed to do so.

Local Private Sector Initiatives

GS Telecom has full VSAT operator licences in five countries (Nigeria, Ghana, Cote d'Ivoire, Tanzania and Mozambique) and has obtained end-user authorisation on behalf of its customers in a further 22 countries. Wilken Afsat is also in the process of building a similar continent-wide service.

Mobile: Sucking up Capacity

Increasingly, mobile operators are carrying more voice traffic than their fixed-line counterparts: by December 2001 Africa had and estimated 30m mobile subscribers and 20m fixed-line subscribers. Where mobile operators possess international gateway licences, they are using satellite to carry international traffic. A key impact of this is that mobile operators are mopping up the excess (C-band) capacity which remains. 'This has disrupted the established pattern' says James Trevelyan of Kingston Inmedia. 'In the past the main users were the PTO and corporate sector, whilst the last three to four months have seen African cellular operators driving demand for capacity'.

Where mobile operators do not have international licences and must switch traffic through a monopoly PTO, that PTO's requirement for international bandwidth has increased and so unless it is able to use fibre this will back up on its requirement for satellite capacity. MTN Nigeria, for example, uses the satellite network of GS Telecom to provide its international connectivity using PanAmSat and New Skies, and also to carry GSM traffic between mobile switching centres (MSCs) in Lagos, Port Harcourt and Abuja.

Mobile operators are also using satellite for their transmission networks: backhauling traffic from remote base stations to central MSCs, or between various MSCs. Indeed, for the time being, most of satellite capacity is used for domestic connectivity rather than international traffic [perhaps on a ratio of 4:1]. Key examples of this are where terrestrial transmission Mattel (Mauritanian-Tunisian Telecommunication Company) in Mauritania where the capital (and MSC) is located at the edge of the network, and in DRCongo where new entrant Vodacom Congo is utilising a satellite backbone; both examples reduce the transmission infrastructure required to span large countries (or those with challenging terrain) to allow faster or more cost-effective roll-out, saving the need to roll-out terrestrial infrastructure. Because it was not able to interconnect to the NITEL transmission, MTN Nigeria has leased a whole transponder on PanAmSat E1.

The greater importance of satellite transmission for mobile is bringing greater engagement by mobile operators into the satellite business. Telecel International, for example, uses M-Link based in Belgium which provides a gateway service for each of Telecel's African subsidiaries that are routing telephone traffic out of their respective countries and delivers traffic from international carriers into Africa using the Telecel networks. Econet Satellite Services (ESS), based in the UK, was set up by Zimbabwean-based operator Econet to operate and market international carrier services using satellite-based infrastructure. ESS also aims to provide routing service for traffic emanating from operations in which EWI acts as operator.

MSI Cellular, the mobile operator with subsidiaries in thirteen African countries, announced in February 2002 that it had acquired the satellite operator LinkAfrica from Intercel. This will 'improve data and voice connectivity within and between these countries, to those areas which cannot be reached by terrestrial solutions and between African countries and the rest of the world.' Celtel DRC, MSI Cellular's subsidiary in the Democratic Republic of Congo, which has become the market leader with 70,000 customers since launching service in December 2000, had been LinkAfrica's largest single customer. Since installing an earth station in Kinshasa in 1992, LinkAfrica has built out a VSAT network covering DRC cities and also Guinea and Madagascar where Telecel has networks, and now rents an entire transponder on the Intelsat 707 satellite to connect international traffic via teleports in the US and Belgium.

ISPs

Except for those countries with fibre, all Internet traffic is satellite-based. ISPs have traditionally been the end-purchasers of bandwidth, usually being forced to procure this from the PTO. However, in one or two countries they are free to link directly to satellites themselves and most often this role has been filled by satellite service providers which lease capacity from the satellite operator and provide the technical services to ISPs. This means that ISPs have had to operate with high fixed costs, which have impeded their ability to penetrate the market on the basis of cost, and so Internet access remains too expensive for vast majority. 'Cost is the throttle on the development of Internet access,' said John Foley of Botswana ISP, referring to bandwidth cost. With Internet exchange points (IXPs) present in only a handful of countries to facilitate peering between ISPs, the bulk of Internet traffic in most countries is international. In a market like South Africa which has peering point, perhaps 70% of Internet traffic is kept local. In a market without such a facility, local traffic accounts for 30%.

The Pan African Virtual Internet eXchange (PAVIX)

This embryonic internet exchange point by being promoted by the African ISP Association (AfrISPA) has the ambition to take the IXP model to a regional level. Each Internet eXchange Point (IXP) will have an earth station and connect via satellite via a point-to-point link. Each ISP will pay for its use of bandwidth through the IXP, but because it is negotiated in bulk discounts will applicable. The intended effect of this is first to remove the need to route traffic through North America or Europe between different African ISPs and different countries. Secondly, acting together, African ISPs will have enough collective traffic to be able to negotiate with carriers to peer traffic with the main Internet carriers, rather than buy bandwidth. This will have the effect of halving costs because will pay for a half link (to peer) rather than a full link (to interconnect via the peering facility in the US or Europe). At this stage, several African ISP Associations are negotiating the agreement to form this PAVIX, but a number of hurdles have yet to be overcome – appointment of the carrier for one, and obtaining the international licensing in countries.

SATELLITE SERVICE PROVIDERS

Between the satellite operators and end-users stand a large range of satellite service providers, operating (and competing) in different sectors, countries and markets. These providers buy capacity from the satellite operators to resell voice, broadcasting and data circuits to a variety of end-users. Increasingly, these traffic on the same Internet protocol (IP) platform: 'The real focus is on IP over satellite' says Martin Jarrold of the Global VSAT Forum, 'when we talk of satellite/VSAT access we must not confine our purview solely to the 'public' Internet, and thereby forget the many other (usually more business orientated) IP-based applications.' The most important innovative uses of satellite/VSAT for Internet access do not result from the deployment of new technology – however fast this is evolving. More important is how service providers are overcoming the two barriers to the use of satellite for Internet access: regulatory restrictions and cost.

Conserving Bandwidth

For current usage, the amount of supply bandwidth may not be insufficient for demand, points out Chris Bell, CEO of Redwing. However, the inefficient, or unnecessary use of bandwidth squanders that which is available. 'You either use it or lose it' said Chris Bell, CEO of Redwing: 'like a plane taking off with empty seats. Once the plane is in the air, you cannot fill the seats. You've paid for it, but you cannot use it.' ISPs, satellite operators and service providers/bandwidth resellers need to work together to concentrate on maximising the use of that capacity which is available. New public-private initiatives are needed to formulate and implement new systems for better bandwidth management, for example, by caching content locally.

Regulations

In many African countries, VSAT is perceived as a threat by the PTT/ PTO and by the regulator. 'While some administrations have progressed quickly, other nations have not realised their full potential, partially because outmoded regulations inhibited or prevented the cost-effective provision of VSAT-based services.' Botswana, for example, has liberalised the use of VSAT and in February 2001 issued three international data gateway licences.

'This is problematic for the service providers who are ready and willing to bring cost-effective connectivity for both the rural unconnected and corporate interests alike. It is also problematic where cross-border connectivity solutions are needed, but are stifled because of what we describe as a 'regulatory honeycomb effect' – with adjacent nations pursuing different policies and maintaining different regulatory frameworks'

Solutions

The African Satellite Corporation, for example, has developed a remote Internet access system without the need for a VSAT licence. Established in 1997 as the INMARSAT service provider for the mining and construction industry in Africa, ASC has INMARSAT satellite phones and also VSAT services including ISDN. This uses INMARSAT uplink and a data cast downlink, and because all countries are signatories to the INMARSAT treaty, the need for authorisation has already been secured.

A second method, used extensively, is using satellite downlink, with landline for the return path. As already seen, this broadband downlink results in the asymmetric nature of Internet traffic. However, latency between the two media means that in practice download times are long. This receive-only satellite licence is not available in many countries. Licensing restrictions in South Africa mean that this is illegal, so the signal must pass through the circuits of a South African company licensed to do so. Presently, this is restricted to Telkom, Orbicom, Transtel and (shortly) Sentech.

Cost

A.key development in VSAT deployment is that newer satellites themselves have much greater power than before. In turn, this means that smaller VSAT antennas can be used to receive signals from them. The impact of this on the market is that smaller (and therefore cheaper) terminals can be used, and with less power required.

PanAmSat is preparing a low-cost VSAT service for sub-Saharan Africa which will lower the cost tenfold, they estimate. This will focus on the lower tier of those smaller users to whom cost is prohibitive in getting Internet access, those 64 Kbps links rather than 512 Kbps – 2 Mbps range, or those 3 Mbps upwards. (In many ways, this reflects the structure of African economies, in which contribution of small, micro and medium-sized enterprises (SMMEs) to GDP is often greater than that of large corporates but little recognised. PanAmSat is acting as a catalyst in working with equipment vendors and service providers to cost structure, and proposing a model. The service provides resellers of capacity DVB as opposed to SCPC so allowing for shared capacity and better bandwidth utilisation.

TRENDS

Sentech, the state-owned South African signal distributor, will from 7 May be licensed to operate an international telecommunications gateway and 'multimedia' services in South Africa. This will allow it to compete in the international segment against Telkom and the second national operator (SNO) when that is licensed. It will act as a carrier of carriers - not be able to retail capacity to end-users itself, but rather wholesale to those service providers which are licensed to do so, be they PSTS licensees or mobile operators.

A trend is seeing the migration of Internet service providers (ISPs) further up the value chain into new segments, from being the end purchasers of bandwidth from providers. A growing number are becoming infrastructure providers themselves, both in provision of local access solutions to reach their customers (typically using fixed wireless access (FWA) or, indeed VSAT) so as to be able to increase their customer base, but also satellite solutions themselves. Some have been using satellite for the downstream to their customer, so as to enable to overcome the bandwidth bottleneck of the last mile and to provide broadband to customers. In South Africa, companies offer this: Siyanda and M-Web. M-Web Broadband delivers to DTH satellite TV dishes, together with the fixed-line dial-up connection for the return path. This is available to M-Web customers in South Africa, although is technically feasible within the Ku-band footprint of the PanAmSat PAS 7 Satellite.

New Breed of Service Provider

We are witnessing the emergence of a new breed of service provider, regional IP operators with the ability to offer voice, data and broadcast and - limited in their ability to offer communication services only by the licensing regime of whichever country they are present in. A key factor for the economic development of African countries is that these commercial operators select those countries as target markets in which they can obtain licences to operate.

For example, SimbaNET, Tanzania ISP, is launching a new service called IP Sys. SimbaNET's IP Sys service creates a direct channel from the Internet backbone to an ISP's local service point, using high-bandwidth Digital Video Broadcast (DVB) or Single Channel per Carrier (SCPC) stream. The coverage area allows the use of single downlink points in any of 20 countries in East, Central or Southern Africa or the creation of multiple downlink points for larger networks. 'We will take networks to the remotest corner of Africa' said Ajit Jatania, a Tanzanian national and MD of SimbaNET, 'We have demonstrated this already by offering services in all regional centres of Tanzania. Why should 85% of the population be ignored which do not live in cities?'

'Of course for each country the regulator will play an important part,' says Jatania. 'Whilst Tanzania realised that communication to regions was important to kick-start the economy, Kenya is still dilly-dallying with the idea and no clear policy is being announced. It is important that regulators give priority to local entrepeneurs' Vice-versa, there is also the entrance into the market of a new breed of operator: regional IP operators. These are providing regional satellite-based connectivity over packet switched networks, and operations on the ground. Transtel, the South African transport parastatal is an example and claims to operate the continent's largest private telecommunications network, and part of the second national operator (SNO) has an international gateway in South Africa. The two most significant are UUNet Africa and Paconet.

UUNet Africa is the joint venture formed between UUNet and Africa Online. Specifically, will connectivity to the Africa Online ISPs in the countries in which it is already present and to those to which it is expanding. In Botswana, for example, UUNet Africa has acquired both an ISP licence and an international gateway licence. UUNet Africa plans to deploy 'Africa's first satellite-based IP backbone.' This is intended to act as an aerial Internet exchange point (IXP) allowing Internet traffic from ISPs to be interconnect with that of others. UUNet Africa are leasing full transponders on strategic satellites to provide this bandwidth; the immediate will be to interconnect the traffic from Africa Online's different operations, but also other ISPs to the network.

Pan African Communications Network (PACONET) is another regional IP operator. It has acquired full international VSAT operating licences in a number of countries. PACONET's vision is to bring low-cost international telecommunications to Africa, and is 'creating a fully meshed African Voice and Data Network with points of presence (PoPs) in all major African cities. These PoPs will carry international telecommunication traffic throughout the continent as well as to and from the rest of the world. A fully operational Pan African Network will allow African nations to communicate directly with each other, rather than having to switch through one of the major US or European hubs. In the Democratic Republic of Congo, Paconet has been able to take this one step further and in addition to licences to operate International VSAT Gateways and roll out Wireless Local Loop networks, it has been granted licences to offer Voice over Internet Protocol (VoIP) services. In one step, provide Internet access and rural telephony services.

The parallels between UUNet Africa's and PACONET's projects and AfrISPA's initiatives are important to note. Can AfrISPA's objectives be achieved more rapidly by working with these commercial projects?

5

Economics of Satellite Communidations

GETTING THE RIGHT BUSINESS MODEL

As the telecommunications industry goes through major transitions of re-regulation, converged services on next generation packet networks, new models of access and metered pricing the satellite industry is 'discovering itself' through experiences such as major and growing successes in DTH entertainment as well as uncertain investments in what now appears to have been highly speculative new mobile satellite telephony projects. So the question remains: what commercial communication satellite businesses and associated architectures will prevail, and why? The measures of success for commercial communication satellite related businesses are the same as for any business – profitability, popularity, value, competitiveness, market share and simplicity. What is absolutely crucial to remember is that the end customer is only motivated by (and will only pay for) the application, not the satellite-tool – so the satellite and earth station technology ONLY exist for the end application. With this in mind, what do we know about historically successful satellite applications and their relationship to tools/functions efficiently provided by satellite technology?

Investment bank C. E. Unterberg Towbin reports the following revenue (both recently, for 1999, and projected, for 2005) for selected satellite industry categorizations in their "Satellite Book':

Mass Market Service:				
• Television (DTH)				
– (1) US	1999:	$6.9 B	2005:	$18.3 B
– (2) Int'l	1999:	$26.1 B	2005:	$49.5 B
• (3) Telephony (MSS)	1999:	$0.4 B	2005:	$7.7 B
• (4) Broadband Services	1999:	$0.2 B	2005:	$6.2 B
Niche Market Services:				
• Radio (DARS)				
– (5) US	1999:	$0.0 B	2005:	$2.0 B

- – (6) Int'l 1999: $0.0 B 2005: $2.0 B
- (7) Messaging (Little LEOs) 1999: $0.5 B 2005: $1.6 B
- (8) Imaging (remote Sensing) 1999: $0.2 B 2005: $1.9 B

Infrastructure Businesses:

- (9) Capacity Leasing (FSS) 1999: $7.7 B 2005: $14.1 B
- (10) Manufacturing 1999: $10.4 B 2005: $10.9 B
- (11) Launch Services (commercial) 1999: $2.6 B 2005: $1.9 B
- Ground Equipment (VSAT and GPS)
  - (12) VSAT 1999: $1.0 B 2005: $2.2 B
  - (13) GPS 1999: $6.1 B 2005: $13.4 B

One of the most important underlying factors is the bandwidth efficiency of an application – a useful aspect of which is revenue scalability with bandwidth. Some applications have a linearly proportional relationship between bandwidth and users (and revenue); others require a certain amount of bandwidth for which the number of users can grow independently. The latter are broadcast applications – and the property that the number of users can grow essentially independent of bandwidth is a powerful economic force.

It is no coincidence that wireless (especially satellite) systems favor broadcast applications. Of course wireless systems may also be favorable for mobile or remote applications.

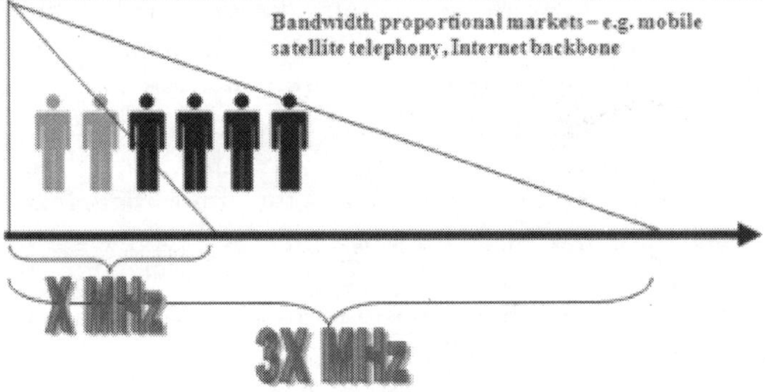

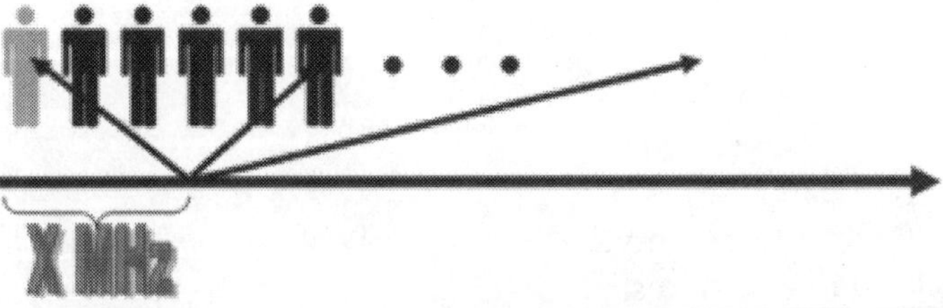

An example of the advantage of broadcast over one-to-one applications, is that a 36 MHz satellite transponder used for Internet access may support somewhere in the range of 5,000 users (a parameter that is not well agreed upon, but is arguably within a factor of 10 of the correct number), whereas the same transponder providing broadcast services together with, say, 29 other transponders may service 10 million (or far more) users. Clearly, an application supporting 333,000+ users per transponder will be highly favored over one that supports only about 5,000 users per transponder.

Which of Unterberg's applications are primarily of broadcast nature? Television, radio and GPS (1, 2, 5, 6 & 13) are the purest broadcast applications from the Unterberg list (GPS satellites broadcast time & location for receivers to use to compute their own location). Telephony, messaging, and imaging (3, 7, 8) are clearly the most non-broadcast applications. Satellite manufacturing and launch services (10 & 11) follow each end application in some proportion. What about Unterberg's other categories – broadband, capacity leasing, and VSAT (4, 9 & 12)? Broadband will surely have some split between broadcast (multicast) and one-to-one applications (although the proportions are yet unknown). Capacity leasing has historically favored (70 to 80%) broadcast applications such as cable & DTH programmers or TV network programmers. The remainder of leased satellite capacity can be broadly categorized as VSAT applications – with both broadcast and one-to-one data flows.

Totaling all broadcast components (including 80% of capacity leasing, and 50% of VSAT figures) comes to $55.8 B in broadcast-related satellite revenue in 1999 versus $3.1 B for non-broadcast applications. It is safe to say that satellites are predominantly a broadcast tool.

1999 Satellite-Related Business Revenues by Broadcast/ Non-Broadcast

		Broadcast	**Non-Broadcast**
1.	US DTH	$6.9 B	
2.	Int'l DTH	$26.1 B	
3.	Telephony (MSS)	$0.4 B	
4.	Broadband Services	?	?
5.	US DARS	$0.0 B	
6.	Int'l DARS	$0.0 B	
7.	Messaging	$0.5 B	
8.	Imaging	$0.2 B	
9.	Capacity Leasing	$6.2 B	$1.5 B
10.	Manufacturing	-	-
11.	Launch Services	-	-
12.	VSAT	$0.5 B	$0.5B
13.	GPS	$6.1 B	
TOTAL		$55.8 B (94.7%)	$3.1 B (5.3%)

Satellite-Related Business Revenue Projections (from Unterberg) by Broadcast/ Non-Broadcast

	Broadcast	Non-Broadcast
1. US DTH	$18.3 B	
2. Int'l DTH	$49.5 B	
3. Telephony (MSS)		$7.7 B
4. Broadband Services	?	?
5. US DARS	$2.0 B	
6. Int'l DARS	$2.0 B	
7. Messaging		$1.6 B
8. Imaging		$1.9 B
9. Capacity Leasing	$11.3 B	$2.8 B
10. Manufacturing	-	-
11. Launch Services	-	-
12. VSAT	$1.1 B	$1.1B
13. GPS	$13.4 B	
TOTAL	$97.6 B (86.6%)	$15.1 B (13.4%)

Unterberg's projections indicate an increase in the share of non-broadcast applications by a factor of 2.5 (5.3% to 13.4%). Such a departure from underlying economics is questionable. The increase derives primarily from Unterberg's projected growth in the telephony (MSS) industry segment – an area where projections have proven to be very wrong in the past.

EMERGING BUSINESSES

What about the applicability of satellites to mobile telephony or broadband Internet?

Mobile Telephony

Mobile telephony via satellite has had less than an auspicious start, and its future is not at all assured. The one-to-one economics of satellite mobile telephony means that the business model is especially sensitive to infrastructure costs – limiting flexibility to change service prices in order to accelerate penetration.

Also important, although to a lesser degree, LEO systems (especially circular-orbit) waste a majority of their orbit time over areas void of customers. Factoring in other aspects, such as singular-application satellite designs, infrastructure threshold for initialization (complete constellations required), replenishment rates, etc. and the risks of mobile satellite telephony service businesses become apparent.

Terrestrial competition has been correctly identified as a significant reason for poor subscriber numbers among early satellite mobile telephony entrants, since broad coverage from terrestrial cellular relegates satellite systems to the most underdeveloped markets (exactly those markets with the fewest customers). The question remains whether satellite mobile telephony business models can be revamped in any way to offer a sufficiently superior solution to terrestrial mobile telephony systems.

SELECTED SATELLITE MOBILE TELEPHONY SYSTEM PARAMETERS

	Iridium	ICO	AceS / Garuda
Average retail service rate per minute	$3	$1.95	< $1
Average wholesale service rate per minute	$1.5 – 2.25	$1 – 3	$0.25 - $0.40
Total Cost	$5.6 B	$4.8 B	$1.3 B
Space Segment Cost	$3.2 B	$2.0 B	$1.2 B
Voice Circuits per Satellite	1,100	4,500	13,750
Number of Satellites	66	10	1
Total Voice Circuits	72,600	45,000	13,750
Coverage Inefficiency (approx.)	90%	90%	30%
Useful Voice Circuits	7,260	4,500	9,625
Total Cost / Useful Voice Circuit	$771,000	$1,070,000	$135,000

Broadband

Broadband Internet via satellite is now spotlighted center-stage, and we have to ask ourselves which of wireless's strengths (broadcast, mobility, remote) will be the driver(s?) for broadband satellite systems. A thoughtful answer to this question should provide some needed guidance to the developers of these new technologies and business plans. As always, first and foremost, satellite's broadcast advantages should be considered.

According to DTT Consulting's report on satellites and the Internet, the majority of satellite Internet circuits today are backbone and access circuits with only a minority of multicasting (broadcasting) circuits. This is consistent with the history of satellites first being used to reach remote areas for one-to-one applications (initially international telephony and data circuits, until fiber optics supplanted the higher-cost satellite circuits), followed by a transition to broadcast (or multicast) applications as new broadcast models develop. The rapidly increasing popularity of using the Internet for both consumer and business applications is certainly a candidate for a new satellite broadcast business model. In the short term, one-to-one satellite circuits are increasing rapidly in advance of wired systems being put into place.

Anticipated prerequisites for a highly-scalable satellite multicast Internet business model are tools that rationalize the Internet's homogeneous one-to-one legacy infrastructure with a heterogeneous satellite multicast extension.

In the meantime there are still many parts of the world that lack terrestrial infrastructure, leaving satellites as an interim one-to-one solution.

While the broadcast application of satellite Internet multicasting holds great promise for long-term differentiation, most satellite Internet links today are one-to-one backbone or access. Several broadband satellite systems have been proposed to address the broadband access/backbone market. Most noteworthy are DirecPC/Spaceway, DISH Network, Teledesic, SkyBridge, Astrolink, and KaStar. The plans for these systems remain somewhat fluid, as they vie to meet a moving-target market.

Company	Backer	Launch	Throughput
DirecPC (/Spaceway)	Hughes Network Systems	Available Now (2002)	400 kbps (16 kbps – 6 Mbps)
DISH Network	EchoStar	2001, tentatively	30 Mbps
Teledesic	Bill Gates, Saudi Prince Talal, Boeing, Motorola	2004	64 Mbps
SkyBridge	Alcatel	Not yet set	N/a
AstroLink	Lockheed Martin, Liberty Media, TRW, Telespazio	2003	20 Mbps
KaStar	Kleiner-Perkins, TV Guide	2002	N/a

The relative economics of advanced broadband systems to traditional satellite systems for Internet access is an interesting point of comparison. A typical broadband GEO satellite may have a coverage region composed of 100 beams, each carrying ¼ of the entire spectrum (to avoid beam-to-beam interference). This leads to an advanced broadband satellite system with 25 times the total system capacity of a traditional system. But the advanced satellite is expected to cost about 2.5 times the cost of the traditional satellite, so the cost effectiveness of each unit of capacity is only improved by a factor of 25/2.5 = 10. But the 100 beams do not uniformly cover the market, and if we estimate the market-match inefficiency as 10%, we arrive at the same cost effectiveness of space segment for advanced broadband systems as for traditional satellite systems. Of course the total capacity for the advanced system is higher (although peak capacity at any high density demand area is lower), and the ground equipment costs are reduced because of more favorable link parameters for spot beams. But total capacity and ground equipment costs are less important than recurring infrastructure costs for satellite Internet access.

Most experts agree that satellite broadband systems will not be able to compete with DSL, cable, or other terrestrial–based infrastructure for one-to-one applications – on account of significantly higher satellite infrastructure costs. This means that satellite broadband systems will be relegated to parts of the world where broadband infrastructure is lacking, which will be where demand density (prospective customers/ square kilometer) is low – somewhat reminiscent of satellite mobile telephony systems. The penetration that can be expected with satellite broadband systems will be lower than what will be achieved with terrestrial systems, according to the higher price that must be charged by satellite-based providers.

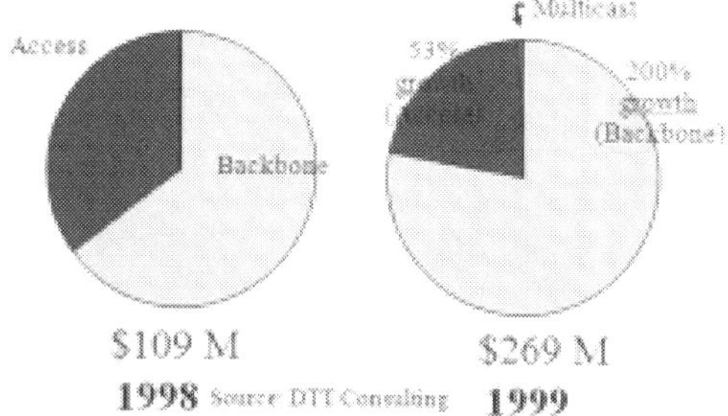

Flying in the face of this logic, however, is a highly optimistic projection by Pioneer consulting for satellite broadband revenue. It is hard to imagine why Pioneer estimates satellite broadband revenue exceeding that of DSL, but Pioneer has certainly gotten our attention that there may be some type of new opportunity here.

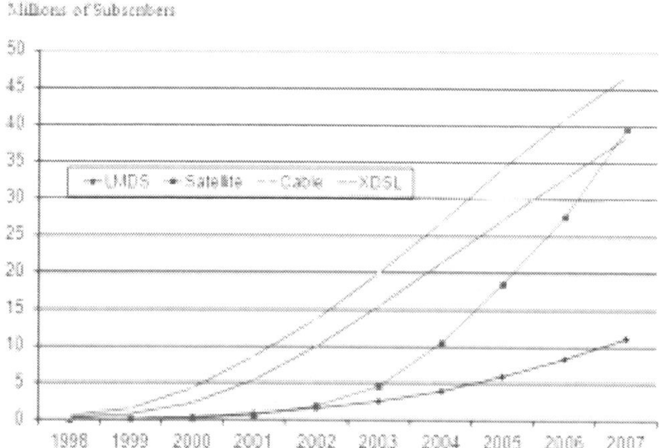

Retail versus Wholesale

Another point of differentiation for satellite business models is the relative position of the business within the distribution architecture, easily illustrated by the different positioning of DTH vs. satellite distribution to cable head-ends. Although both businesses leverage satellite's broadcast advantage, the leveraging has a markedly different characteristic for each business model. Total U.S. DTH annual revenue is about $6.9 billion over about 12 million subscribers – or $575/ subscriber/ year. While total U.S. cable annual revenue is about $33 billion over about 68 million subscribers, the revenue collected by satellite transponder leasing to cable programmers is equivalent to only about $250 million/ year, or only about $3.70/ subscriber/ year (only about 0.6% of the revenue per sub of DTH). Of course, the scope of DTH businesses is much broader than transponder leasing business, and profitability is not addressed by this comparison. The important difference is that in the DTH case, the DTH service provider benefits from the broadcast nature of satellites, AND generates revenue proportional to the number of subscribers, whereas in the transponder leasing case, the leasing company (satellite services provider) benefits from the broadcast nature of satellites (because the provider leases capacity to broadcast programmers), BUT DOES NOT generate revenue proportional to the number of subscribers or number of cable head-ends.

OTHER FACTORS

Economies of scale. The greatest opportunity for volume cost reduction is in receive-only satellite terminals, but there are also efforts underway by satellite operators/service providers to standardize two-way VSATs (the DVB – SIT, digital video broadcasting satellite interactive terminal standard) in order to lower costs and consequently accelerate adoption of two-way services. Volume cost reductions from standardized components has greatest impact when the number of components is in the range of 100's of thousands or millions – a number not achieved for typical space-borne units, but achievable in some cases for ground station equipment..

Regulations. International and national telecommunications regulations from the World Trade Organization, International Telecommunications Union, and national regulators, continue to rebalance the scale between competitive opportunities and protectionism. A trend towards technology-independent regulation and the possibility of further regulations over the Internet each has the potential to change the nature of satellite-based businesses. Technology innovations. Innovations in satellite technology offer the opportunity to differentiate products and obtain higher revenue/profitability. Opportunities to differentiate with proprietary technology advancements are present not only in the basic satellite platform and traditional broadcast application areas, but especially in Internet interconnectivity with satellites, local-station distribution, certain types of on-board processing, and interactivity in general.

GIS INDUSTRY OVERVIEW: BUSINESS SEGMENTS AND OPPORTUNITIES

DATA SIDE

The most important component of Geographic Information Systems is its requirement for spatial data. Spatial data is any kind of information that has been collected, compiled, or processed with a spatial component, that is, a tie to a geographic location on the surface of the Earth. It so happens that this is a large segment of the spatial industry, often consuming an appreciable portion of dollars assigned to GIS implementation projects. Spatial data management is increasingly a consideration in any information management system (IMS) due to the fact that large amounts of data are being collected with spatial components. Businesses and government organizations are realizing that a traditional IMS does not allow an organization to leverage the value of spatial information inherent in their data. This has led to the development of software tools as extensions to commercial Data Management Systems (DMS) that allow for better storage, manipulation, and query of spatial data.

Considering the importance of quality spatial data, there are a number of operations that play important roles in the GIS Industry:

Data Collection and Extraction

There are a great number of sources for spatial data and, thereby, a large variety of collection systems. Organizations involved in primary data development utilize a number of technologies to gather unique, and typically current, spatial data. These may include aerial photography and satellite imagery, specialized sensors and metering systems such as stream flow monitors, human surveys by the Census Bureau, or customer information collected by supermarkets and businesses.

The fact that spatial information is collected does not necessarily entail its usability in a GIS. There is often some level of extraction or processing of relevant information to get data into a desired format or category. For example, satellite imagery provides an exceptional source of information for current conditions at a given location on the surface of the earth. If the topic of analysis is transportation related it may be necessary to extract a transportation network, roads, railroads, and canals, from the imagery. This is not a simple or easy process and is prone to significant quantitative and qualitative error. A noteworthy problem in this area is a lack of national and international standards for this process.

This issue is being addressed however. There are a great number of government and commercial organizations that specialize in the collection and processing of primary spatial data. Aerial photography has been collected, by the USGS for example, for decades and there is a substantial infrastructure in place dealing with this data source. High resolution satellite imagery as collected by Space Imaging, L.P., of Thornton Colorado, and a number of other firms in the near future, will provide a great deal of data for international sites that were previously unavailable on the open market. This provides a significant international market opportunity, especially in support of international Nongovernmental Organizations (NGO's).

In short, the US market has been well represented by quality spatial data while much of the international market has been significantly impoverished.

Data Conversion and Compilation

New collection systems are not the only source for spatial data. There is a great volume of data existing in map libraries and historical data sets and archives that are utilized in Geographic Information Systems. They provide a valuable secondary source of spatial information and a significant sector of the spatial information market. The most relevant example of such sources is the USGS 7.5 minute quadrangle maps covering the US. This is a systematically compiled, high quality map collection that provides a valuable source of base map information for a broad selection of GIS application areas. Attractive aspects of the data set are the uniform system of collection, established standards of quality, and regular and systematic process for updating.

As with satellite imagery, the fact of the existence of such data sets does not mean they are immediately useable in a GIS. Paper maps must be converted to a digital format that is then populated with appropriate descriptive or attribute information before their value can be realized. This is referred to in the industry as digitizing and is a highly labor intensive process as few quality automated systems have been developed for this operation. Digitizing existing data sets is essential to getting existing information into a format that allows for spatial analysis and query in a digital environment. Again, there are a great number of quality issues involved in this process. The fact that the map base was systematically compiled is a significant benefit.

There is a great volume of spatial data that is currently available in a digital format. This does not necessarily indicate that it is immediately available for use by organizations in the pursuit of their tasks. Data may come from a variety of sources and in a variety of formats such that a great deal of processing is required to convert and compile it into a useable form. For instance, a customer information system may include data fields recording street address information. If it is desired to analyze customer characteristics or habits in a spatial context there is a need for converting that street address, typically a text string, into a point on a map. This requires a highly accurate street network with comprehensive addressing information. The problem can be further complicated if spatial data from other sources is going to be sought and synthesized. Most implementations of GIS at an organizational scale involve a great deal of effort in collecting and compiling a useable spatial data set. The requirements and sources will depend on the tasks at hand and existence of suitable information. Again, this process involves a well-established commercial base and significant efforts by the federal government. As older datasets are converted and newer ones are compiled in digital and spatially enabled formats this segment of the industry will decline in relevance.

Data Management and Integration

Spatial information comes from a variety of sources, in a broad range of often-proprietary formats and is contained in a number of frequently incompatible data management systems. This is particularly the case in government organizations that have historically had specialized systems developed for dealing with information they are mandated to collect and maintain. Due to resource limitations and the rate of advances in technology they are left with a plethora of data sets maintained in obsolete and incompatible specialized systems. Given the vast sources, formats, and systems for dealing with spatial information, this leads to significant needs in the field of data management and systems integration.

Databases must be established that maintain the spatial integrity of data. Data from different sources often requires processing to create a cohesive data set that coincides spatially and minimizes errors in coordinate systems or conversions. The establishment of SOP's for adding to or changing foundation data is vital to maintaining a quality data implementation.

SYSTEM SIDE

While becoming less of a concern considering the competency of modern IMS and the interconnectedness of modern computer platforms, the system side of GIS is still a very significant issue. GIS has specific hardware and software requirements, out of the box most commercial GIS applications will interface with only selected IMS. Spatial data have unique issues related to their use and management that contribute to the issue of system implementation. A broad integration of GIS into an organizational setting needs to take these issues into consideration along with the special circumstances of the programs seeking support.

Organizational Integration and Training

Government, at all levels, and commercial businesses collect and manage information that has spatial components. Typically, this information is not maintained in a spatially enabled manner and is thereby not utilized efficiently or effectively in decision-making processes. All levels of government; local, state or federal, can use GIS technology in virtually every department or agency. This is due to the fact that the topics of their attention are related to jurisdiction and government duties. Some applications may be intuitive, such as natural resource management and landuse planning, since cartographic work has played a role in these areas since their inception. Less intuitive but increasingly important uses of GIS are found in the realm of Law Enforcement, Health Services, Insurance Analysis, Emergency Planning and Response... the list continues.

The major hurdle to broad and integrated application of GIS in these organizations is a lack of understanding with regard to what GIS is, how it can be utilized, and what benefit/roll it can play in an organization. A major task lies in the area of demonstrating the utility and cost effectiveness of implementing broadly applied and well integrated systems. Typically, there are examples of well integrated GIS operations across the country to use as case studies and exemplars.

As an example, Loudon County Virginia, until fairly recently a rural county in the Washington DC suburbs, was an early-adopter of GIS technology. In the mid 1980's the county government, to aid in development planning and land records management, implemented a GIS program. At this time it is recognized as one of the top county level GIS operations in the country. Data is managed and maintained by a central office with 14 different county departments drawing on and contributing to the data system. This has truly been a great benefit to the county as it has ranked in the 3rd and 8th place in the list of fastest growing counties in 1999 and 1998 respectively. Having the system in place has greatly aided the management of rapid urbanization from issues of where to build new schools to what neighborhoods are going to complain about the location of a proposed telecommunications monopole. Organizations that are not yet using GIS but could benefit from it need insight to how it can be applied. This involves understanding the operations of the organization, the types of issues or problems they deal with, and the customers they serve. A great deal of communication and training is necessary to place the "pretty map" technology in the context of powerful decision support tool.

Infrastructure Implementation

This leads to the issue of system composition. A GIS operation may consist of a single laptop used for fieldwork, a number of workstations networked together, a multi-level pan-organizational implementation in government, or an international operation monitoring climate data. The possibilities and permutations are as varied as the topics to which GIS can be applied. Essential equipment and software will be dictated by the tasks at hand, as will staffing needs, and data requirements. The key, as with any technology implementation, is getting the systems to work together effectively and in a manner that maximizes system availability and data integrity. Modern platforms are much more conducive to a networked and integrated environment while legacy systems may not be. This is particularly an issue with government operations.

Organizations aiming to implement or expand a GIS operation require guidance on which systems to maintain, which to migrate, and which to replace or rebuild from the ground up. Obviously the resources available and the budget at hand will dictate a major part of this decision process.

Software Application Development

As GIS is introduced to new topic areas and expands its base in existing areas of application, there is an increasing need for specialized software development. Most commercial GIS vendors provide a development environment for their software products that allow for the creation of specialized software tools using GIS functionality and targeted at specific research or operational goals. The intent of such efforts is to provide a user a simplified set of tools that allow them to perform their jobs without exposing them to the complications of spatial data management and analysis.

This specialized tool development has resulted in GIS software applications that are designed for use against very specific issues such as Crime Analysis and Mapping, Precision Agriculture, Logistics Operations, or Marketing and Business Analysis. A government agency may desire the development of a GIS based tool that allows citizens to identify the location of hazardous waste sights in their local region. The average citizen does not need to know how the information is processed or stored; they are concerned with the answer. A specialized tool would be developed to perform the operation with a simplified interface and reporting system to minimize confusion on the part of the citizen. The advent of the World Wide Web has moved a portion of this market to the Internet.

A significant portion of work in the GIS field involves this type of application development for specialized use and accessing or analyzing unique data sets. GIS vendors are largely moving to open development systems with specialized object sets that will allow the inclusion of spatial analysis tools with common development platforms such as Visual Basic, C++, and Java.

Information Query and Distribution

One of the major goals of spatial data management and GIS integration is to leverage the spatial component of information to improve the decision making process. Many applications of GIS involve the enabling of data in the spatial sense, allowing for improved questioning and better answers. In most instances the information desired by a user will only be a subset of that available.

In this context a GIS query may consist of extracting a subset of a database, defined by spatial parameters, that will then be incorporated into a project. This involves the development and management of a system that allows a user to select specific parameters or data sets, identify the information they require and have it delivered to them as a package. Again, the Internet provides an ideal medium for performing this type of operation.

An application of GIS that perfectly demonstrates such an operation is the development of a GIS enabled real estate system. This ideally would allow a real estate buyer to establish a set of criteria; house price, school system, restaurants, public transportation, recreational facilities, etc., that would then provide them a list of available property in a given area including maps of locations, house photos, and other relevant information aiding them in making a decision. The technology for such an operation exists; the major reason it is not in operation is due to resistance on the part of the industry.

A number of government organizations are developing mechanisms for distributing data in their care. The USGS provides comprehensive data sets for the entire country via Web-based query tools. The data is not yet provided in an integrated fashion and requires a respectable amount of knowledge and skill on the part of the user for operational exploitation. The data and its intended use plays a major roll in the complexity of the system, query mechanisms, and distribution avenues required.

THE APPLICATIONS SIDE

We have discussed a number of segments in the GIS industry that are integral to its operation and growth. In the process many examples of how GIS are applied, or "Applications of GIS" (not in the software sense) have been mentioned and discussed. GIS is a very flexible tool that is relevant to innumerable issues.

A less than exhaustive list:

- Advertising
- Air Traffic
- Agriculture
- Architecture
- Automated Mapping
- Banking
- Business
- Cadastral/Tax Mapping
- Census
- Community Development
- Construction
- Crime Analysis
- Defense
- Demographics
- Direct Marketing
- Education
- Emergency Services
- Engineering
- Environmental Management
- Epidemiology
- Facility Management
- Financial Services
- Fleet Management
- Forestry
- Health Care
- Hotel Marketing
- Insurance
- Intelligence

- Land Management
- Local Government
- Logistics
- Manufacturing
- Military/Defense
- Natural Resources
- Oil and Gas
- Pipeline
- Property Management
- Public Health
- Public Information
- Public Safety
- Public Transit
- Publishing
- Real Estate
- Redistricting
- Retail Siting
- Route Planning
- Target Marketing
- Tax Assessment
- Telecommunications
- Tourism
- Transportation
- Travel
- Trucking
- Utilities
- Water/Wastewater
- Wildlife Management

This remains a very important aspect of the industry that has not changed, and will not, despite technological advances. GIS has its foundation in very specific, project-based applications and many of the uses of GIS are still project or topically based.

Project Based, Applied GIS

GIS provides a powerful tool for considering problems or issues with spatial components. This leads to its use on a project-by-project basis that aims to address and answer specific questions. As an example, when considering water quality issues for a given region or watershed, the Chesapeake Bay perhaps, a GIS may be developed to deal with that region or specific issue. The only data used is that which covers the area of interest, questions asked and analysis performed are relevant to the topic and issues being investigated with the goal of achieving some answers. Once those are determined to a satisfactory level the project may be retired and a new region or issue may be addressed. This will continue to be an important sector of the industry but often requires a variety of specialized skills and knowledge on the part of the organization performing the analysis. Fundamental knowledge of GIS is essential to the process but a detailed knowledge of the topic being investigated is equally important. Finding professionals with both is ideal but not always possible. This is especially true when reaching into new areas of GIS use and application. Developing a close relationship with those portions of an organization involved in the thematic area is essential.

Research and Development

GIS is a relatively new and burgeoning field. It has grown rapidly along with other sectors of the technology industry and this trend will continue with an increasing integration of spatial data, processes, and functionality. The industry is not yet mature and there remains a great deal of problems with the science of the technology to be worked out. For example, temporal issues and time as a continuum are not easily dealt with in a GIS. Most systems deal with time in the context of time slices or snapshots, in much the same way that maps provide a single slice in time, that are outdated once collected. Much of the research work on GIS issues is relegated to university research centers and fall into the emerging discipline of Geographic Information Science (also known as GIS).

CONCEPT OF SATELLITE COMMUNICATIONS

Scientists from different countries conceived various ideas for communications through space along with the technological breakthroughs in different fields of science. The Russian scientist Konstantin Tsiolkovsky (1857-1935) was the first person to study space travel as a science and in 1879 formulated his Rocket Equation, which is still used in the design of modern rockets. He also wrote the first theoretical description of a man- made satellite and noted the existence of a geosynchronous orbit. But he did not identify any practical applications of geosynchronous orbit. The noted German Scientist and rocket expert, Hermann Oberth, in 1923 proposed that the crews of orbiting rockets could communicate with remote regions on earth by signaling with mirrors.

In 1928, Austrian Scientist Hermann Noordung suggested that the geostationary orbit might be a good location for manned space vehicle. Russian Scientists in 1937 suggested that television images could be relayed by bouncing them off the space vehicles. During 1942-1943, a series of articles by George O Smith were published in Astounding Science Fictions concerning an artificial planet, Venus Equilateral, which functioned as relay station between Venus and Earth Station when direct communication was blocked by Sun. However, Arthur C. Clarke, an electronic engineer and the well-known science fiction writer is generally credited with originating the modern concept of Satellite Communications. In 1945, Clarke, in his article `Extra Terrestrial Relays: Can Rocket Stations give Worldwide Radio Coverage?' published in Wireless World outlined the basic technical considerations involved in the concept of satellite communications. Clarke proposed orbiting space stations, which could be provided with receiving and transmitting equipment and could act as a repeater to relay transmission between any two points of the hemisphere beneath.

He calculated that at an orbital radius of 42,000 km. the space station's orbit would coincide with the earth's rotation on its axis and the space station would remain fixed as seen from any point on the earth. He also pointed out that three such synchronous stations located 120 degrees apart above the equator could provide worldwide communications coverage. The concept was later considered to be generating a billion dollar business in the area of communications. However, Clarke did not patent the most commercially viable idea of twentieth century as he thought satellites would not be technically and economically viable until the next century.

REALISATION OF CONCEPT TO REALITY

In October 1957, the first artificial satellite Sputnik -I was launched by former Soviet Russia in the earth's orbit and in 1963 Clark's idea became a reality when the first geosynchronous satellite SYNCOM was successfully launched by NASA. The realization of the concept of satellite communications from an idea to reality has been possible due to a large number of technological breakthroughs and practical realization of devices and systems, which took place during and after the World War II. The pressures of international military rivalry during cold war period were also able to a great extent to push scientific and technological research and development far faster than it would have been possible if applied for peaceful purposes.

The successful launching of communications satellite in earth's orbit was possible because of keen interests shown by specific groups of people along with the developments in diverse areas of science and technology. Some of these factors, which are considered important in the realization of satellite communications, are:

- Development of high power rocket technology and propulsion systems capable of delivering satellites in high altitude orbits
- Scientific and military interests in Space Research
- Development of Transistors and miniaturization of electronic circuitry.
- Development of Solar Cells for providing sustained energy source for the satellite.
- Development of high-speed computers for calculating and tracking orbits.
- Government support in large-scale financial commitment to Space

Technology Development for Military, Scientific Experiments and Civilian Applications.

- International military rivalry among super powers.
- The psychological impact of Sputnik Challenge leading to long range program of scientific research and development undertaken by US.

Before the transformation of the concept of communications by satellite to blue print and subsequent development of the hardware took place it was necessary to make the scientific communities convinced about the technical feasibility of such a system.

In US J.R. Pierce, of Bell Laboratories initiated this by promoting the idea of transoceanic satellite communications within the scientific and technical communities. In 1955 Pierce in a paper entitled Orbital Radio Relays proposed detailed technical plan for passive communications satellites, disregarding the feasibility of constructing and placing satellites in orbit. He proposed three types of repeaters.

- Spheres at low altitudes
- A plane reflector
- An active repeater in 24 Hr. orbit

Pierce concluded his paper with a request to the scientific community to develop rockets capable of launching communications satellite. Fortunately, scientific and military interest in rocketry after World War II contributed in the development of a number of rockets like Atlas, Jupiter and Thor rockets in US and different multistage rockets in former USSR that ultimately made the launching of satellites in orbit possible.

On Oct. 4, 1957, Sputnik-1 was launched as part of Russia's program for International Geophysical Year. The launching of Sputnik marks the dawn of the space age and the world's introduction to artificial satellite. Mass of Sputnik was only 184 lbs. in an orbit of 560 miles above the earth. It carried two radio transmitters at 20.005 MHz and 40.002 MHz.

However this space craft was far more than a scientific and technical achievement as it had a tremendous psychological and political impact particularly on United States resulting in a technological competition between United States and Russia, long term planning in Space Research and establishment of NASA. Four months after the launch of Sputnik, US Explorer-1 was launched in January 1958 by a Jupiter rocket and the space race between Russia and US began.

EVOLUTION OF COMMUNICATION SATELLITES

During early 1950s, both passive and active satellites were considered for the purpose of communications over a large distance. Passive satellites though successfully used in the early years of satellite communications, with the advancement in technology active satellites have completely replaced the passive satellites.

Passive Satellites

The principle of communication by passive satellite is based on the properties of scattering of electromagnetic waves from different surface areas. Thus an electromagnetic wave incident on a passive satellite is scattered back towards the earth and a receiving station can receive the scattered wave. The passive satellites used in the early years of satellite communications were both artificial as well as natural. In 1954, the US Naval Research Laboratory successfully transmitted the first voice message through space by using the Moon to scatter radio signal. These experiments resulted in the development of Moon-Relay System, which became operational in 1959 for communications between Washington, DC and Hawaii and remained operational till 1963.

The first artificial passive satellite Echo-I of NASA was launched in August 1960. Echo-I was 100-ft. diameter inflatable plastic balloon with aluminum coating that reflected radio signals transmitted from huge earth station antennas. Echo-I had an orbital height of 1000 miles. Earth Stations across US and Europe picked up the signal and contributed a lot in motivating research in communication satellite. Echo-I was followed by Echo-II in 1964. With Echo-II, Scientists of US and Soviet Russia collaborated for the first time on international space experiments. Signals were transmitted between University of Manchester for NASA and Gorki State University in Russia. The orbit of Echo-II was 600 to 800 miles.

In 1963, US Air Force under Project West Ford launched an orbital belt of small needles at 2000 miles height to act as a passive radio reflector. Speech in digitized form was transmitted intelligently via this belt of needles. However, further work in this area was discontinued due to strong protests from the astronomers.

Although passive satellites were simple, the communications between two distant places were successfully demonstrated only after overcoming many technical problems. The large attenuation of the signal while traveling the large distance between the transmitter and the receiver via the satellite was one of the most serious problems. The disadvantages of passive satellites for communications are:

- Earth Stations required high power (10 kW) to transmit signals strong enough to produce an adequate return echo.
- Large Earth Stations with tracking facilities were expensive.
- Communications via the Moon is limited by simultaneous visibility of the Moon by both the transmit and the receive stations along with the larger distance required to be covered compared to that of closer to earth satellite.
- A global system would have required a large number of passive satellites accessed randomly by different users.
- Control of satellites not possible from ground.

Active Satellites

In active satellites, which amplify and retransmit the signal from the earth have several advantages over the passive satellites. The advantages of active satellites are:

- Require lower power earth station
- Less costly
- Not open to random use
- Directly controlled by operators from ground.

Disadvantages of active satellites are:

- Disruption of service due to failure of electronics components on-board the satellites
- Requirement of on-board power supply
- Requirement of larger and powerful rockets to launch heavier satellites in orbit

World's first active satellite SCORE (Satellite Communication by Orbiting Relay Equipment) was launched by US Airforce in 1958 at orbital height of 110 to 900 miles. It transmitted a pre-recorded message of Christmas Greetings from US President Eisenhower. However, the satellite did not function as a true repeater.

The first fully active satellite was Courier launched into an orbit of 600 - 700 mile, by Department of Defense in 1960. It accepted and stored upto 360,000 Teletype words as it passed overhead and rebroadcast them to ground station farther along its orbit. It operated with 3 watts of on-board output power and it was also the first satellite to use solar cells for generating electrical power.

In July 1962 AT&T's active satellite Telstar was developed and launched. Telstar was placed in an elliptical orbit with orbital height of 682 to 4030 miles circling the earth in 2 hours and 40 min. Through Telstar, the first live transatlantic television was transmitted. Voice, television, fax and data were transmitted between various sites in UK, France, Brazil Italy and US at 6/4 GHz frequency range. Relay-I satellite of RCA & NASA, was the first satellite to carry redundant system for increasing the reliability. Telephone & Television signals were transmitted to Europe, South America and Japan. Frequency bands of 4.2/1.7 GHz and orbit heights of 942 to 5303 miles were used.

Syncom, the first geosynchronous satellite of NASA was built by Hughes Aircraft Co. and was launched in July 1963 and was used for conducting many experiments. Most famous of the series Syncom-III was launched in 1963 and was used to transmit Tokyo Olympic games to United States, demonstrating the commercial market for space technology. Syncom-I and-II were used by Department of Defense for military purpose. The Syncom Satellites marked a turning point in the development of Satellite Communications as most of the commercial satellites that followed were designed to operate from geosynchronous orbit. Table gives the major milestones of Space Radio Communications events, prior to the start of commercial satellite communications service by INTELSAT.

Table. Major Milestones of Space Radio Communications

Category	Year	Activity	Person/Agency/ Country.
Geostationary concept	1945	Suggestion of Geostationary satellite communication feasibility.	A. Clark (U.K)
Moon Reflection	1946	Detection of Lunar Echo by Radar	J. Mofenson (U.S.A.)
	1954	Passive relaying of voice by moon reflection.	J.H. Trexler (U.S.A.)
	1960	Hawaii-Washington, D.C. Communication by Moon Reflection.	U.S.A. Navy.
Low altitude orbit.	1957	Observation of signals from Sputnik -1 Satellite.	U.S.S.R., Japan and others.
	1958	Tape-recorded voice transmission by Satellite SCORE.	U.S.A. Air Force.
	1960	Meteorological facsimile Trans mission by Satellite Tiros-1.	U.S.A. NASA
	1960	Passive relaying of telephone and television by Satellite Echo-1.	U.S.A. Army.
	1960	Delayed relaying of recorded voice by Satellite Courier-1B.	U.S.A. Army.
	1962	Active transatlantic relaying of communication by Satellite Telstar-1.	U.S.A., U.K., France.
	1962	Communication between manned Satellites Vostok-3 and 4; Space television transmission.	U.S.S.R.
	1963	Scatter communication by tiny needles in Orbit. (West Ford Project 6)	U.S.A. MIT.
	1963	Active transpacific relaying of communication by Satellite Relay 1.	U.S.A. NASA, Japan.
Synchronous Satellite.	1963	USA-Europe-Africa communication by Satellite Syncom 2.	U.S.A. NASA
	1964	Olympic Games television relaying by Satellite Syncom 3	U.S.A., NASA Japan.
	1965	Commercial Communication (Semi-experimental) by Satellite Early Bird.	Inter Sat.

SATELLITE COMMUNICATIONS SYSTEMS

Historically, commercial operational satellite communications systems were developed after having working experience with a large number of experimental satellite systems launched to demonstrate the various aspects of satellite communications. Initially the commercial satellite communications systems were meant for meeting the needs of international transoceanic communications. The trend was for establishing only a few large earth stations in any country for overseas communications. In the early years of satellite communications, the earth stations were large due to low transmit power available from the satellites.

Over the years the trend has been reversed as with the advancement of technology, higher transmitted power is available from the satellites. This has reduced the size and the cost of the earth stations. Thus the trend is now on the use of thousands of small earth stations and portable hand held terminals, for meeting the various specialized communications needs. Moreover, apart from international system a number of Regional, Domestic, and military systems are now in operations worldwide. From the traffic point of view, emphasis was initially more on point-to-point telephone, telex etc, and to some extent on Television broadcasting. The present trends however are on Direct To Home television broadcasting and VSAT based data communications using small antenna systems deployed on rooftops or on one's backyards. Finally, it is expected that the satellite communications will meet the ultimate goal of hand held personal communications of voice and data for anyone from anywhere and anytime.

Different types of Satellite Communications Systems are:

- Experimental
- International
- Regional
- Domestic
- Military
- Navigational and Radio Determination
- Personal Communications System
- Broadband Satellite System

Experimental Satellite Communications Systems

For the purpose of test and evaluation of new technologies a number of satellites have been designed and operated for technical experiments. Various experiments have also been conducted using these satellites for demonstrating different applications of communications satellites. Prominent among these experimental satellites are:

- Applications Technology Satellite Series (ATS-1, ATS-3, ATS-5 & ATS-6) of NASA.

- Joint Canadian - US Communications Technology Satellite (CTS or Hermes)
- Advanced Communications Technology Satellite (ACTS) of NASA.
- APPLE (Ariane Passenger Payload Experiment) Satellite of India.
- Symphonie Satellite (France & Germany).
- SIRIO (Italy)
- LES (US military)
- OTS (ESA)
- JBS, CS (Japan)

International Satellite Systems

Currently only a part of the world's long distance telecom traffic is handled by different international satellite communications systems. However, for international broadcasting of television there is no alternative to satellite communications. Examples of various international satellite systems are:

- Intersat
- New Skies Satellites
- PanAmSaT
- Intersputnik
- Inmarsat
- Cospas-Sarsat

Interlsat: Recognizing that Satellite Communications would be an important means for international cooperation, in July 1961, President Kennedy of US invited all nations to participate in a communication satellite system in the interest of world peace and brotherhood among peoples throughout the world.

In Dec. 1961, UN endorsed the US proposal regarding the desirability of a global system of communication satellites because it could

- Forge new bonds of mutual knowledge and understanding between people everywhere
- Offer a powerful tool to improve literacy and education in developing areas
- Support world weather services by speedy transmittal of data
- Enable leaders of nations to talk face to face on a convenient and reliable basis

The UN unanimously adopted General Assembly Resolution, which stated that: Communications by means of satellite should be available to the nations of the world as soon as practicable on a global and nondiscriminatory basis'. In August 1962, US Government passed Communications Satellite Act. Its purpose was to establish a commercial communications system utilizing satellites, which would serve the needs of the US and other countries and contribute to world peace and understanding. The significance of the choice of a single system for international communications is economic, technical and political. In August 1964, the final negotiations for the international satellite system were completed and nineteen nations became the founding members of International Telecommunications Satellite Organization (INTELSAT) with Headquarters in Washington D.C, USA. These nineteen nations are Australia, Austria, Belgium, Canada, Denmark, France, Germany, Ireland, Italy, Japan, the Netherlands, Norway, Portugal, Spain, Sweden, Switzerland, the United Kingdom, United States and Vatican City. Over the years the number of member governments grew to 144.

In April 1965, Early Bird (INTELSAT-I) was launched starting the commercial international satellite services. Within four years the INTELSAT system grew from the single transatlantic link to the global network with high capacity INTELSAT satellites positioned over the Atlantic, Pacific and Indian Ocean Regions. The 240 voice circuit capacity of the Early Bird is miniscule in comparison to the channel capacity of the latest INTELSAT satellites which caters to tens of thousands of telephone channels in addition to providing TV, data, fax, telex and Internet services to more than 200 countries and territories. With the improvement in life of the satellites, introduction of latest communication techniques, and the availability of more channel capacity, the tariff of Intelsat has also been reduced considerably over the years.

In the 1960's at the time of establishment of Intelsat, the satellite Communication Industry was not well developed. The international telecommunications was also not considered suitable for handling by private companies. However, the skepticism changed after successful privatization of telecommunications departments in many countries during the last few decades of the twentieth century. Since in highly competitive telecommunications market, private enterprises are in a position to provide better and cheaper services compared to the international organizational set up of Intelsat, ideas for privatization of Intelsat were mooted. Privatization places Intelsat on a level playing field to better address opportunities of the telecommunications marketplace. Streamlined decision-making is expected to make it easier to expand the business and introduce new services. Considering these, in November 2000, the representatives of all member governments of Intelsat unanimously approved a plan to privatize Intelsat.

The approved plan endorses the transfer of all assets, liabilities and operations to a private Bermuda based company known as Intelsat Ltd., and its 100 % subsidiaries. In accordance with its heritage as a global satellite communications services provider to all countries, Intelsat Ltd. will continue to honour a clear set of public service commitments on a commercial basis. These include

- Global coverage and global connectivity
- Service to "lifeline" customers around the world according to specific Lifeline Connectivity Obligation Contracts
- Non-discriminatory access to the Intelsat Ltd. Satellite fleet

A small separate and independent inter governmental office will monitor the private company's implementation of these public service commitments. Privatization of INTELSAT is expected to be completed in 2001. New Skies Satellites N.V (New Skies) is formed through the partial privatization of Intelsat. It is a wholly independent satellite service provider starting its services through five in orbit satellites transferred from Intelsat fleet. New Skies, with headquarters at The Hague, Netherlands, began operations as a commercial spin off from Intelsat in November 1998, with three satellites in Atlantic Ocean Region, one each in the Indian & Pacific Ocean Region and appropriate ground facilities around the world. These satellite and the ground facilities provide complete global coverage at C band and high powered Ku band spot beams over most of the world's principal population centers. It offers video, voice, data and Internet communications links for broadcast networks, telephone carriers, enterprise customers and ISPs.

PanAmSat Corporation, based in Greenwich, Conn./ USA, is a leading provider of global video and data broadcasting services via satellites. The company builds, owns and operated networks that deliver entertainment and information to cable television systems, TV broadcast affiliates, telecommunications companies and corporations with a large fleet of 21 satellites in orbit as on 2001.

Intersputnik: The Intersputnik International Organization of Space Communications with headquarters at Moscow was established in 1971, according to an intergovernmental agreement submitted to UN, for operating a global satellite communication system. However, Intersputnik became operational only in 1974 with Molniya and Statsionar Satellites for providing telephony, telegraph, radio, data, telex and direct broadcasting of television. Initially Intersputnik had nine member countries, which has grown to twenty-three over the years. These are Afghanistan, Belarus, Bulgaria, Cuba, Czech Republic, Georgia, Germany, Hungary, Kazakhstan, Kyrghizstan, DPR Korea, Laos, Mongolia, Nicaragua, Poland, Tajikistan, Turkmenistan, Romania, Russia, Syria, Ukraine, Yemen, Vietnam. Intersputnik's user base exceeds 100 state run and private companies.

Due to the changes in the geopolitical conditions in the 1990's and rapid development of telecommunications market with the introduction of new services created severe competitions among the telecom service providers. Under the changed circumstances, Intersputnik reviewed its strategy to get adapted to the dynamically developing environment. In order to keep the competitive edge, Intersputnik established strategic alliances with different satellite communications operators, manufactures, launch vehicle service providers, ground equipment manufacturers and international entities. One of these alliances is the joint venture Lockheed Martin Intersputnik (LMI) established in 1997. First LMI satellite with 44 high power C and Ku band transponders and lifetime of 15 years was launched at 75 deg E. in September

1999. With strategic alliance with Lockheed Martin corporation, Intersputnik is able to provide high quality services, which include digital video, high bit rate access to Internet, use of VSATs, Telemedicine, Tele-education, banking on a global scale etc.

Inmarsat: Inmarsat was established as an international cooperative organization similar to Inmarsat, for providing satellite communications for ships and offshore industries. Inmarsat, a specialized agency of UN, was established in 1979 and became operational in 1982 as a maritime focused intergovernmental organization with headquarters located at London. Inmarsat has forty-four members and also provides services to nonmember countries. Inmarsat has become a limited company since 1999. INMARSAT Ltd is a subsidiary of the Inmarsat Ventures plc holding company, which operates a constellation of geosynchronous satellites for worldwide mobile communications. The satellites are controlled from Inmarsat headquarters at London, which is also home to Inmarsat ventures and a small Intergovernmental office created to supervise the company's public service duties for the maritime community i.e. Global Maritime Distress and Safety System and Air traffic Control communications for the aviation industry.

Starting with a user base of about 900 ships in early 1980's, the user base of Inmarsat grew to 210,000 ships, vehicles, aircrafts and portable terminals in 2001. Inmarsat Type A mobile terminals meant for installation in large ships are quite expensive, whereas, portable INMARSAT mini-M terminals are small, cost effective and easy to operate. The services provided by INMARSAT include telephone, fax and data communications up to 64 kbps. Other services include videotext, navigation, weather information and Search & Rescue. Inmarsat Satellites can also be used for emergency Land Mobile Communications for relief work and to re-establish communications or to provide basic service where there is no alternative. Inmarsat can also be used to alert people on shore for coordination of rescue activities. Apart from maritime and Land Mobile Satellite Service, Inmarsat also provide aeronautical satellite service for passenger communications.

Inmarsat system operates at C-band and L-band frequencies. The INMARSAT system uses allocations in the 6 GHz band for the ground station to satellite link, 1.5 GHz for satellite terminal downlink, 1.6 GHz for terminal to satellite uplink and 4 GHz for the satellite to ground station down link.

COSPAS-SARSAT: COSPAS-SARSAT is a joint venture of Canada, France, Russia and US. It is a satellite based international Search and Rescue, alert and location system using different low earth orbit satellites operating in the frequency of 120 MHz and 406 MHz. Local Users Terminals operating in different parts of the world pick-up the alert signals from Emergency Location Beacons carried by ships and airplanes and pass on the information regarding the location of accident to the nearest rescue centres for carrying out the rescue operations. Since the operationalisation of the satellite based Search and Rescue

System, hundreds of lives have been saved due to timely deployment of prompt rescue operations.

Domestic Satellite Communications Systems

In the initial years of implementation of commercial Satellite Communications the emphasis was mainly on the transoceanic and international communications. However use of satellite communications for improvement of domestic communications also emerged as a distinct possibility. Countries like, USSR, Canada, Indonesia took initiatives in implementing domestic satellite communications systems for the respective countries. Satellite Communications System for domestic communications is cost effective compared to the conventional terrestrial systems under the following conditions.

- A large country without basic terrestrial communications
- Population is spread over mountains, deserts and a large number of islands
- Thinly populated remote areas

Former USSR was the first country to adopt satellite communications for its domestic use. However, because of its geographic location, where large landmass was in the high latitude region, the geosynchronous satellite systems were not found to be suitable. Thus a system with a series of Molniya Satellites operating in highly elliptical non-geosynchronous inclined orbits was introduced in 1965 to meet the country's domestic requirement of telecommunications and television transmission. Canada became the first country to use a geosynchronous satellite for domestic communications with the launching of Anik-1 in 1972. With the advent of Anik Satellite it was possible to cover for the first time the whole of Canada particularly thinly populated northern region under the live TV coverage. Apart from TV, Anik Satellites are also used for radio broadcasting to remote locations and interactive distance education.

Indonesia is the first developing country to have its own domestic satellite system. Because of its limited infrastructure and widely scattered population dispersed over more than 13,600 islands, a satellite communications system is an ideal technology to deliver telecommunications and broadcasting throughout the country. Telecommunications services using PALAPA-A, the first Indonesian domestic satellite started in 1976. Some of the other countries with their own domestic satellite communications systems are:

- United States of America (Wester, SBS, Etc.)
- India (INSAT)
- Brazil (Brazilsat)
- Mexico (Morelos)
- China (Chinasat)
- Japan (CS, BS)

Many countries where operating an exclusive domestic satellite communications system is not economical, the domestic requirements of communications can be met by leasing capacity from Intelsat or other satellites.

Regional Satellite Communications Systems

Regional Satellite Communications Systems have been an ideal means to deliver telecommunications and broadcasting services to a number of countries in a region for meeting their domestic and regional telecommunications and broadcasting requirements rather than having separate domestic system for each of these countries. A number of regional satellite communications systems are presently in operations and quite a few of them are in the planning stages.

Some of the regional Satellite Communications Systems are:

- Eutelsat
- Arabsat
- Aussat
- Palapa

Eutelsat is a consortium of twenty-six European nations, established for operations and maintenance of space segment of the Eutelsat Satellite System, and providing its members with the space segment capacity necessary for meeting their telecommunications services requirements.

Eutelsat provides services that are not available via Intelsat in Europe. These include:

- Transmission of television networks to eighteen countries for cable distribution and transmission of Eurovision programs
- Intra-European telephony and telegraphy
- Multi-service data communications for computer networking, facsimile, remote printing of newspapers, teleconferencing etc.

Eutelsats' space segment is coordinated by the European Space Agency (ESA) which procures spacecraft from European manufactures.

Arabsat evolved from 1953 Arab league agreement to develop regional telephone, telex and telegraph communications. Arab Space Communication Organization was established in 1976 and had twenty-two members. However, by the time two Arabsat Satellites were launched in 1985, Egypt, a leader in the System's planning had been expelled from Arab league. Each Arabsat Satellite has twenty-five C-band transponders and one C/S band transponder for community television reception.

However, the Arabsat Satellites are extremely under utilized as only six countries have earth stations. Several countries are still working with Intelsat lease rather than switch to Arabsat. AUSSAT systems of Australia designed for meeting domestic communications needs of Australia for Radio, TV broadcasting and long distance Telephony is also used by Papua New Guinea for telecommunications and broadcasting services. Since 1979, PALAPA system of Indonesia became a regional satellite system, after Philippines, Thailand and Malaysia signed agreement to use PALAPA.

Military Satellite System

For military communications Army, Air force and Navy use both fixed and mobile satellite systems. In addition to the normal communications, military communications are also required for tactical communications from remote and inhospitable locations.

The special requirements of military communication terminals are high reliability, ruggedness, compact, operations under hostile environment, immunity to jamming, ease of portability and transportation, etc. Examples of military satellite communications systems are:

- Dscs (US AF)
- Skynet (UK)
- Nato (Nato)
- Fltsatcom (US Navy)
- Milstar

Because of the special frequency band used in Military satellite system and other special requirements, Military satellite Systems are always much costlier and it takes longer time to design and develop compared to commercial satellite communications systems. Realizing that not all communications are strategic in nature, there is a trend now to use commercial communications system as far as possible. US Department of Defense is one of the major users of commercial Iridium satellite system with their own gateway.

Navigational System

Satellites have now replaced the stars and terrestrial systems for the purpose of navigation and radiolocation. The Transit Satellite system of US Navy was the first satellite navigational system with satellites orbiting in low polar orbits. By means of triangulation, the crews could establish the location of the ship and submarine by picking up the signals transmitted by different Transit satellites. US agreed to allow civilian use of Transit Navigational System for use by merchant marine shipping industry throughout the world. Transit system is now replaced by the Global Positioning System (GPS) of US Navstar Satellite System consisting of eighteen low earth orbiting satellites operating at L-band. GPS receiver calculates the position (latitude, longitude, height) with extremely high accuracy by receiving signals from at least three-satellite passes. Apart from its use in ships, the miniaturized GPS receiver has also found many applications related to land based fleet monitoring. The Russian Glonass system is the other navigational satellite system. However, the system is not being maintained properly by timely replacement of the satellites.

PERSONAL COMMUNICATIONS SYSTEM

Introduction of terrestrial personal communications in many countries since1980's saw a very rapid growth of wireless telecommunications. Considering that the terrestrial cellular telephone systems are limited only to urban and sub-urban areas, the business potential in providing global satellite based personal communications system was realized by many. It was thought that a satellite based personal communication system would provide not only communications to remote locations from anywhere, it would also provide a seamless roaming system integrating the scattered pockets of terrestrial system.

A large number of global and regional personal satellite communications systems using both geosynchronous and non-geosynchronous satellites were planned during the 1990's. Most of these proposed systems were for voice communications with non-geosynchronous satellites in order to avoid the long delay associated with geosynchronous satellites. However, quite a few of these proposed systems, never took off and a few ran into financial problems at the implementation stage casting serious doubts about the commercial viability of satellite based personal communications system using non geosynchronous satellites. Examples of regional satellite based personal communications systems providing voice and data services through geosynchronous satellites are

- Thuraya
- Asia Cellular Satellite (ACeS)

Both these systems use satellites with large antenna systems and cover a large area of Asia and Europe.

Thuraya: Thuraya Satellite Company is a regional satellite system that provides satellite telephone services to a region covering 99 countries through a dynamic mobile phone that combines satellite, GSM cellular system and GPS. Thuraya was established in 1997 in UAE as a private joint venture with shareholders from 18 national telecommunications operators and investment houses. Thuraya meets the demand for seamless coverage of mobile communications to 2.3 billion people residing in India, Middle East,Central Asia, North & central Africa and Europe. Thuraya handsets offer voice, data, fax messaging and position location. It enables the user to use GSM service in local networks and automatically switch on to satellite mode whenever out of local terrestrial reach. The first Thuraya satellite operating in L band has been launched in Oct 2000 and commercial service started from 2001.

ACeS is another regional geo-mobile personal satellite communications system providing digital voice, fax and data communications using small dual mode (satellite and GSM) handsets. Users of ACeS are able to roam between terrestrial GSM cellular and satellite networks and can interface with public switched telephone networks. ACeS is jointly owned by PT Pacifik Satelit

Nusantara of Indonesia, Lockheed Martin Global Telecommunications of USA, Philippines Long Distance Telephone Company and Jasmine International of Thailand. ACeS coverage area extends from Pakistan & India in the west to Philippines & Papua New Guinea in the east and from China & Japan in the north to Indonesia in the south. ACeS started its operations from November 2000.

Examples of a few Personal Satellite communications Systems providing services or planning to provide services using non-geosynchronous satellite constellations are:

- Iridium (66 Satellites)
- Globalstar (48 satellites)
- Orbcomm (35 satellites)
- New ICO (10 satellites)

Non-geosynchronous satellites in low and medium earth orbits need a large number of satellites in a constellation to provide global coverage and the number of satellites in the constellation increases with decreasing orbital height. The non-geosynchronous low earth orbit satellites appear to be attractive for providing two-way voice and data communications and location positioning to small handheld terminals from the points of view of higher available power from the satellite, low time delay etc. However, launching and maintaining a large number of satellites on orbit and operations of corresponding ground system pose technical as well as operational problems. Financial problems faced by a few of these systems in providing services at the early stages of operations, have made people rethinking on the commercial viability of such systems. The frequency allocations for personal communications systems are in the VHF, L band and S band.

Iridium: The Iridium system was the first satellite based Personal communications system to start commercial global wireless digital voice communications operations in November 1998 with its 66 Low Earth Orbit satellite constellation. But the infamous original Iridium service did not pick up and it failed in its attempt to attract the target subscriber base. This caused financial problems and bankruptcy of the company within a few months of starting the operational service. Iridium's failure despite a sophisticated on board technology and compatibility of hand sets with different terrestrial mobile telephone standards, is considered largely due to poor marketing and a service that was too costly. Moreover, by the time Iridium system was launched the cellular phone coverage also improved worldwide, thus reducing the target service area of Iridium that was uncovered by terrestrial cellular service.

After acquiring the assets of the bankrupt Iridium LLC, a privately held corporation Iridium Satellite LLC, launched its commercial global satellite communications service in March 2001. Iridium Satellite provides voice, paging, and messaging services to mobile subscribers using handheld user terminals. Frequency of operations of Iridium system is L band. Globalstar: Globalstar is a consortium of leading international telecommunications companies originally established in 1991 to deliver satellite telephony services through a network of exclusive service providers. Globalstar system designed with a constellation of 48 low-earth orbiting satellites, started its commercial phone service using multimode handsets from October 1999. Other Globalstar services include voice mail, short messaging service, fax and supporting terrestrial IS 41 and GSM systems.

Calls from a Globalstar wireless handsets are transmitted in L band to the satellite and the receive frequency is in S band. Calls via satellites are routed through the appropriate gateway, from where they are passed on to existing fixed and cellular telephone network. The service is available in more than 100 countries in 6 continents. Orbcomm: Orbcomm Global LP is the first commercial provider of global low earth orbit satellite data and messaging communications system. Globalstar started its commercial service in November 1998 with 28 out of a constellation of 35 low earth orbit satellites and 14 gateway Earth stations in 5 countries. Orbcomm provides two-way monitoring, tracking and messaging services to both fixed and mobile terminals. The system is capable of sending and receiving two-way alphanumeric packets, similar to two-way paging and e-mail.

VHF frequency bands are used for providing two-way messaging services at low data rates. The orbiting satellites pick up small data packets from sensors in vehicles, containers, vessels or remote fixed sites and relay these to the destination through a tracking Earth Station and Gateway Control Centre.

Orbcomm was originally formed as a partnership company owned by Orbital Sciences Corp (USA), Teleglobe Inc (Canada) and Technology Resources Industries Bhd (Malaysia). In April 2001, International Licensees, a consortium of Orbcomm licensees and other investors purchased all the assets of Orbcomm Global LP and its other entities that were under protection of bankruptcy since September 2000.

New ICO: New ICO, formerly of ICO global communications is working on a Medium Earth Orbit(MEO)/ Intermediate Circular Orbit (ICO) satellite system designed for both fixed and mobile operations around the world. ICO Global Communications was founded in 1995 and contracts for satellites launch services and ICO network were awarded. In August 1999, due to financial problems, the company was declared bankrupt. However, with fresh investment from a group of international investors New ICO emerged from the bankruptcy. New ICO system consists of a constellation of 10 on-orbit ICO satellites, 2 on orbit spares at an orbit of 10,390 km. Target launch of service of New ICO system is 2003.

New ICO is based in London with offices in different countries. The goal of New ICO is to provide global Internet protocol services, including Internet connectivity, data, voice and fax services. The system operates in both circuit switched mode based on GSM standard and packet switched Internet protocol mode. New ICO plans to target markets like maritime, transportation Government, oil, gas, construction & other industries, individuals and small & medium size businesses.

Broadband Satellite System

Broadband satellite service is an emerging service which has caught the fancy of many for meeting the demand of worldwide fiber like access to telecommunications services such as computer networking, broadband Internet access, interactive multimedia and high quality voice. These systems use advanced satellite technology at Ka band or Ku band frequencies to achieve the high bandwidth requirements.

Examples of proposed Broadband Satellite systems are:

- Teledesic
- SkyBridge
- Spaceway

Teledesic satellite network is designed with 288 plus spare satellites at low earth orbits. The operating frequency is in the Ka band of the frequency spectrum with 30 GHz uplink and 20 GHz downlink. The network will enable millions of simultaneous users to access the two-way network using standard user equipment providing up to 64 Mbps on the down link and up to 2 Mbps on the up link. The fixed user equipment will be mounted out door and connect inside to a computer network or PC.

Teledesic is a private company based in Bellevue, Washington (USA) attracting investment from many reputed companies and individuals. The ambitious Teledesic service targeted to begin in 2005 will enable broadband connectivity for businesses, schools and individuals everywhere on the planet and expected to facilitate improvements in education, healthcare and other crucial global issues.

SkyBridge is a satellite-based broadband global telecommunications system designed to provide business and residential users with interactive multimedia applications as well as LAN interconnection or ISDN applications, thus allowing services such as high speed Internet access and video conferencing to take place anywhere in the world. The system is based on a constellation of 80 low earth orbiting satellites, which link professional and residential users equipped with low cost terminals and terrestrial gateways. The satellite network operates at Ku band and will deliver asymmetric broadband connection to fixed network at up to 60 Mbps (in steps of 16 kbps) to the user and up to 2 Mbps (in increments of 16 kbps) on the return link via a gateway.

SkyBridge LP was formed in Delaware, USA in 1997. The partners of SkyBridge LP are Alcatel and leading industries from North America, Europe and Asia. Spaceway is another advanced broadband satellite system offered by Hughes Network system of USA that will make high-speed broadband applications available on demand to the businesses and to consumers around the world. Operating in the Ka band spectrum, SPACEWAY will consist of interconnected regional satellite systems providing service to nearly all of the world's population. The first North American regional service will start in 2002 with two geosynchronous satellites plus an on orbit spare. Using a globally deployed system of satellites in conjunction with a ground-based infrastructure, users will transmit and receive video, audio, multimedia and other digital data at uplink rates between 16 kbps to 16 Mbps. The access to the system will be provided through a family of low cost easily installed 66 cm terminals.

ORBITS FOR COMMUNICATION SATELLITE

The path a Satellite or a planet follows around a planet or a star is defined as an orbit. In general the shape of an orbit of a satellite is an ellipse with the planet located at one of the two foci of the ellipse. The circular orbit may also be considered as an ellipse where the two foci of the ellipse coincide at the center of the circle. Satellite Orbits are classified in two broad categories i.e.

Non-Geostationary Orbit (NGSO)

Early ventures with satellite communications used satellites in Non-geostationary low earth orbits due to the technical limitations of the launch vehicles in placing satellites in higher orbits. With the advancement of launch vehicles and satellite technologies, once the Geo Stationary Orbit (GSO) was achieved, majority of the satellites for telecommunications started using GSO due to its many advantages.

During 1990s the interests in NGSOs were rekindled due to several advantages of NGSO in providing global personal communications in spite of its many disadvantages.

Advantages of NGSO are:

- Less booster power required to put a satellite in lower orbit
- Less space loss for signal propagation at lower altitudes (<10,000 km) leading to lower on board power requirement
- Less delay in transmission path – reduced problem of echo in voice communications
- Suitability for providing service at higher latitude
- Lower cost to build and launch satellites at NGSO
- Use of VHF and UHF frequency bands at NGSO permits low cost antennas for hand-held terminals

Disadvantages of NGSO are:

- Requirement of a large number of orbiting satellites for global coverage as each low earth orbit satellite covers a small portion of the earth's surface for a short time.

- Complex hand over problem of transferring signal from one satellite to another
- Less expected life of satellites at NGSO requires more frequent replacement of satellites compared to satellite in GSO
- Compensation of Doppler shift is necessary
- Satellites at NGSO undergoes eclipse several times a day necessitates the requirement of robust on board battery system for the satellite for operations without solar power during eclipse
- Complex network management for a constellation of satellites and corresponding ground system
- Problem of increasing space debris in the outer space

There are different types of Non Geostationary Orbits (NGSO), depending on the orbital height and the inclination of the orbital plane. Inclination is the angle that the orbital plane makes with the equatorial plane at the time of crossing the equator moving from south to north of the earth and is measured from 0 to 180 degrees. NGSOs are classified in the following three types as per the inclinations of the orbital plane

- Polar Orbit
- Equatorial Orbit
- Inclined Orbit

In polar orbit the satellite moves from pole to pole and the inclination is equal to 90 degrees. In equatorial orbit the orbital plane lies in the equatorial plane of the earth and the inclination is zero or very small. All orbits other than polar orbit and equatorial orbit are called inclined orbit.

A satellite orbit with inclination of less than 90 degrees is called a prograde orbit. The satellite in prograde orbit moves in the same direction as the rotation of the earth on its axis. Satellite orbit with inclination of more than 90 degrees is called retrograde orbit when the satellite moves in a direction opposite to the rotational motion of the earth. Orbits of almost all communication satellites are prograde orbits, as it takes less propellant to achieve the final velocity of the satellite in prograde orbit by taking advantage of the earth's rotational speed.

Example of retrograde orbit is the sun synchronous orbit where the orbital parameters are such that that the satellite crosses the same latitude at the same local time. This type of orbit is used for earth observation satellites where repeated observations are required to be made under the same sun angle. It needs more propellant to launch a satellite in retrograde orbit as it is launched in a direction opposite to the direction of the earth's rotation.

Satellite orbits are also classified in terms of the orbital height. These are:

- Low Earth Orbit (LEO)
- Medium Earth Orbit (MEO)/ Intermediate Circular Orbit (ICO)
- Highly Elliptical Orbit (HEO)
- Geosynchronous Earth Orbit (GEO)

Satellite orbits with orbital height of approximately 1000 km or less are known as Low Earth Orbit (LEO). LEOs tend to be in general circular in shape. Satellite orbits with orbital heights of typically in the range of 5000 km to about 25,000 km are known as Medium Earth Orbit (MEO)/ Intermediate Circular orbit (ICO). MEO and ICO are often used synonymously, but MEO classification is not restricted to circular orbits. Satellites in Highly Elliptical Orbit (HEO) are suitable for communications in the higher latitudes. Russian Molnya satellites have highly inclined elliptical orbits with a perigee of about 1000 km, apogee of 40,000 km, inclination of 63.435 deg and orbital period of 12 hours. In Geosynchronous Earth Orbit (GEO) the satellite is in equatorial circular orbit with an altitude of 35,786 km and orbital period of 24 hours. Three satellites in GEO placed 1200 apart over equator cover most of the world for communications purposes.

Geostationary Orbit (GSO)

There is only one geostationary orbit possible around the earth, lying on the earth's equatorial plane and the satellite orbiting at the same speed as the rotational speed of the earth on its axis. For a Satellite to have an orbital period equal to that of earth's rotation i.e. a sidereal day (23 Hrs 56 min. 4.09 sec.) an altitude of 35,786 km is required. Such a satellite orbiting at a velocity of 3.075 km/sec remains fixed relative to any point on earth or geostationary. With the idealized assumptions that the geostationary satellite is at rest relative to the earth the conditions required to be satisfied for geostationary orbit are:

- The orbit shall be circular
- The period of the orbit shall be equal to the period of rotation of the earth about itself
- The plane of the orbit shall be the same as the equatorial plane but the sub-satellite longitude, i.e. the longitude of the projection of the satellite on the Earth's surface can be selected arbitrarily.

The principle of satellite communications based on this concept of geostationary orbit was originated by Arthur C Clarke. Main advantage of geostationary satellite being the permanent contact between the ground segment and the satellite with fixed directional antennas at both the earth station and the satellite. The ITU (International Telecommunications Union), recognizing the importance of the GSO along with the frequency spectrum as limited natural resources available on the earth, set out the procedures for all radio communications services, regarding the use of GSO/spectrum through ITU Radio Regulations, a binding international treaty. With respect to the use of the GSO and frequency spectrum, the ITU space regulations laid down in the ITU Constitution is as follows:

In using frequency bands for radio services, Member states shall bear in mind that radio frequencies and any associated orbits, including the geostationary-satellite orbit, are limited natural resources and they must be used rationally, efficiently and economically, in conformity with the provisions of Radio Regulations, so that countries or groups of countries may have equitable access to those orbits and frequencies, taking into account the special needs of developing countries and the geographical situation of particular countries.

Table below outlines the salient features, advantages and disadvantages of Geostationary Satellite Orbit (GSO).

Table. Geostationary Satellite Orbit

Attitude	35,786 km.
Period	23 Hr. 56 min. 4.091 sec. (One sidereal day)
Orbit inclination.	0^{0}
Velocity	3.075 km per sec.
Coverage	42.5% of earth's surface.
Sub satellite point	On equator.
Area of no coverage	Beyond 81^{0} North and South latitude. (77° if angle of elevation below 5° are eliminated)
Advantages	- Simple ground station tracking.
	- No hand over problem
	- Nearly constant range
	- Very small doppler shift
Disadvantages	- Transmission delay of the order of 250 msec.
	- Large free space loss
	- No polar coverage

A perfect geostationary orbit is a mathematical abstraction that could be achieved only by a spacecraft orbiting around a perfectly symmetric earth and no other forces are acting on the spacecraft other than the central gravitational attraction from the earth. The abstraction is however, useful as an approximate description of real case, since all other forces or perturbations due to attractive forces of the Moon and the Sun and the non-sphericity of the Earth's gravity are small.

In real life due to gravitational pull of the Moon & the Sun, the equatorial orbital plane of the satellite makes an angle of inclination with respect to the equatorial orbital plane. For a satellite with orbital period equal to a sidereal day and non-zero inclination, the footprint of the satellite will move in North-South direction over its sub satellite point instead of remaining stationary. The non-spherical shape of the earth also causes movement of the satellite in the east-west direction. Thus the trace of the satellite on earth appears to roam in both North-South and East-West direction around the sub-satellite point.

The inclination of the satellite can be corrected by firing appropriate thrusters on-board the satellite and is known as North-South station keeping. Similarly the correction of East-West drift of the satellite is called East-West Station keeping. Without any station keeping the inclination plane drifts to about 0.86 deg per year. Thus the satellite orbital position is required to be corrected periodically to keep the drift from the desired location within a certain limit. Considering the drift in the satellite position in North-South and East-West direction around the sub-satellite point, it is more appropriate to designate such an orbit as geosynchronous orbit.

GEOSYNCHRONOUS COMMUNICATION SATELLITE

Geosynchronous Satellites have now become almost synonymous for communications satellites, because of its wide use in telecommunications due to the advantages over non-geosynchronous satellites. Because of the availability of a number of communication satellites over the geosynchronous arc, the communications between different parts of the world have become possible and affordable. The communication satellites have played a significant role in converting the world into a global village.

Salient features of Geosynchronous Communications Satellite

Salient features of Geosynchronous Satellite are:

- Wide Coverage
- Stationary Position
- Multiple Access
- Suitability for transcontinental telecommunications, broadcasting, mobile and thin route communications.
- Frequency reuse capability
- Very low Doppler Shift
- Reliability.
- Cost effectiveness.

Brief description of each of these features are given below:

Wide Coverage: From the geosynchronous orbit the satellite can cover an area equal to about 42% of the area of the earth (38% if angles of elevation below 5º are not used). Thus three satellites placed 120º apart can cover almost the whole world for the purpose of communications. INTELSAT Satellites strategically placed over Atlantic Ocean Region (AOR), Indian Ocean Region (IOR) and Pacific Ocean Region (POR) covers the whole world for International Telecommunications. With worldwide satellite TV coverage, any incidence happening in any part of the world can now be viewed live in the TV throughout the world.

Stationary Position: The orbital velocity of the geosynchronous satellite being equal to the rotational velocity of the earth on its own axis, the satellite in the geosynchronous orbit appears to be stationary with respect to any location from the earth. Thus the satellite is always visible from any earth station situated in its coverage region and the tracking of the satellite is simple and there is no hand over problem of transferring signal from one satellite to another as in the case of satellites in NGSO. The constant visibility of the satellite also enables both the satellite and the earth station to use highly directive antennas. High gain of the antennas on-board the satellite and the earth station, enhances the transmit and receive capabilities.

Multiple Access: Multiple Access is the ability of a large number of users to simultaneously interconnect their respective voice, data and television links through a satellite. The wide geographic coverage and broadcast nature of satellite channel are exploited by means of multiple access. Multiple access also helps in optimum use of satellite capacity, satellite power, spectrum utilization and interconnectivity among different users at reduced cost. A satellite in geosynchronous orbit can link multiple earth stations within its coverage area and separated by great circle distances up to 17,000 Km. Multiple access is the unique feature of satellite communications not possible to get by any other means. For m earth stations visible from a Satellite, the number of potential available communication circuits is given by,

$$n = m\,(m-1)/2$$

compared to non flexible 2-port network of conventional cable or land based networks.

Suitability for Transcontinental Telecommunications, Broadcasting, Mobile and Thin Route Communications: TV Broadcasting via Satellite is perhaps the most common use of geosynchronous satellite. In developing countries where the terrestrial TV distribution is very limited, the communications satellites can be very effectively utilized for TV distribution. Geosynchronous satellites handle a large portion of transcontinental telecommunications traffic. Geosynchronous Satellites along with other NGSO satellites are found to be suitable for reliable mobile communications for ships and aircrafts, as the ship and the aircraft can continuously maintain the communication link with the satellites while moving. However, GEO based satellite systems are much simpler to operate and maintain compared to other system.

Geosynchronous Satellites are also the most suitable means of providing reliable and cost effective communications to thin route rural areas, interconnecting small islands, and providing communications to hilly and difficult terrain. Frequency Reuse: The frequency bands of a geosynchronous satellite can be reused by different methods for increasing the channel capacities of the communications satellite. By using specially designed spatially separated shaped beams the same frequency and polarizations can be reused. By using orthogonal polarizations the same frequency bands can be reused for the same coverage area of the satellite. By using orthogonal linear and circular polarizations and shaped beams covering different regions, the same frequency band can be reused many folds thus increasing the communication capacity of geosynchronous satellite. Different techniques of frequency reuse of the same frequency band are found in INTELSAT series of satellite. Very Low Doppler Shift: Compared to low earth orbit satellites, in geosynchronous satellite there is almost no Doppler Shift i.e. change in the apparent frequency of operations to and from Satellite, caused by the relative motion of the Satellite and the earth station. Satellites in elliptical orbits have different Doppler shifts for different earth stations and this increases the complexities of the receivers especially when a large number of earth stations intercommunicate.

Reliability: The reliability of long distance telecommunication links improves considerably when geosynchronous satellites are used. The path loss in the satellite links although very high; these remain almost constant, thus maintaining the performance quality of the link.

Cost effectiveness: The geosynchronous satellite because of its long life of twelve to fifteen years and wide-band operations shared by a large number of users, makes the point to point service very cost effective compared to the service provided by land based terrestrial system. No viable alternatives to geosynchronous satellites are presently available, so far as the broadcasting and mobile services are concerned.

Problems of Geosynchronous Satellite Communications Systems

The problems of geosynchronous satellite communications systems are:

- No coverage of polar region.
- Long time delay.
- Echo.
- Eclipse due to the earth and the sun.
- Sun Transit outage

No Coverage Region: The geosynchronous satellite from its location of 35,786-Km altitude above equator is not found suitable for communications beyond the latitude of 81 deg. Thus the polar region of the earth cannot be properly covered by geosynchronous satellite.

Time Delay: In Satellite Communications System using geosynchronous satellite, the signal has to travel a long distance while travelling from the transmit earth station to the receive earth station via satellite. From the geometry of the geosynchronous satellite orbit it is found that the single hop time required for the signal to travel from one point to another varies from 230 m sec. (90 deg. elevation) to 278 m sec. (0 deg. elevation). This time delay does not pose any problem in data and broadcasting services, but this delay is quite perceptible in two-way telephone conversations. ITU-T specifies a delay of less than 400 msec to prevent echo effects and delay variation of upto 3 msec.

Although the propagation and intersatellite delays of LEOs are lower, LEO systems exhibit high delay variation due to connection handovers, satellite and orbital dynamics and adoptive routing. Echo: Generally a long distance telephone circuit is accompanied by echo due to mismatch at the terminal point where circuits are converted from four wire to two wire system. As the delay of the echo is increased, the effect of the echo becomes increasingly disagreeable to the talker. The echo can be attenuated by using echo suppressor or echo canceller. By using echo suppressor of excellent quality, a two hop satellite link can be utilized for practical communications, provided the delay is acceptable. Eclipse of Satellite: A Satellite is said to be in eclipse when the positions of the earth, the Sun and the Satellite are such that the earth prevents sun light from reaching the satellite i.e when the satellite is in the shadow of the earth.

For geosynchronous satellites, eclipses occur for 46 days around equinox (March 21 and September 23). During full eclipse, a satellite does not receive any power from solar array and it must operate entirely from batteries. In case the power available from battery is not enough, some of the transponders may be required to be shut down during the eclipse period. The satellite passes through severe thermal stress during its passage into and out of the earth's shadow. The solar power also fluctuates sharply at the beginning and end of an eclipse. For these reasons the probability of failure of satellite is more during eclipse than at any other time. Sun Transit Outage: Sun transit outage takes place when the sun passes through the beam of an earth station. During vernal and autumnal equinox, the sun approaches toward a geosynchronous satellite as seen from an earth station and this increases the receiver noise level of the earth station very significantly and prevent normal operations. This effect is predictable and can cause outage for as much as 10 min. a day for several days. The sun transit outage is about 0.02 percent in an average year. A receiving earth station cannot do anything about it except wait for the sun to move out of the main lobe.

ELEMENTS OF SATELLITE COMMUNICATIONS SYSTEM

Two major elements of Satellite Communications Systems are

- Space Segment
- Ground Segment

The Space Segment includes

- Satellite
- Means for launching satellite
- Satellite control centre for station keeping of the satellite

The functions of the ground segment are to transmit the signal to the satellite and receive the signal from the satellite. The ground segment consists of

- Earth Stations
- Rear Ward Communication links
- User terminals and interfaces
- Network control centre

SPACE SEGMENT

Communication Satellite

Communication satellites are very complex and extremely expensive to procure & launch. The communication satellites are now designed for 12 to 15 years of life during which the communication capability of the satellite earns revenue, to recover the initial and operating costs. Since the satellite has to operate over a long period out in the space the subsystems of the satellite are required to be very reliable. Major subsystems of a satellite are:

- Satellite Bus Subsystems
- Satellite Payloads

Satellite Bus subsystems:

- Mechanical structure
- Attitude and orbit control system
- Propulsion System
- Electrical Power System
- Tracking Telemetry and Command System
- Thermal Control System

Satellite Payloads

- Communication transponders
- Communication Antennas

Since the communications capacity earns revenue, the satellite must carry as many communications channels as possible. However, the large communications channel capacity requires large electrical power from large solar arrays and battery, resulting in large mass and volume. Putting a heavy satellite in geosynchronous orbit being very expensive, it is logical to keep the size and mass of the satellite small. Lightweight material optimally designed to carry the load and withstand vibration & large temperature cycles are selected for the structure of the satellite.

Attitude and orbit control system maintains the orbital location of the satellite and controls the attitude of the satellite by using different sensors and firing small thrusters located in different sides of the satellite. Liquid fuel and oxidizer are carried in the satellite as part of the propulsion system for firing the thrusters in order to maintain the satellite attitude and orbit. The amount of fuel and oxidizer carried by the satellite also determines the effective life of te satellite. The electrical power in the satellite is derived mainly from the solar cells. The power is used by the communications payloads and also by all other electrical subsystems in the satellite for house keeping. Rechargeable battery is used for supplying electrical power during ellipse of the satellite.

Telemetry, Tracking and Command system of the satellite works along with its counterparts located in the satellite control earth station. The telemetry system collects data from sensors on board the satellite and sends these data via telemetry link to the satellite control centre which monitors the health of the satellite. Tracking and ranging system located in the earth station provides the information related to the range and location of the satellite in its orbit. The command system is used for switching on/off of different subsystems in the satellite based on the telemetry and tracking data. The thermal control system maintains the temperature of different parts of the satellite within the operating temperature limits and thus protects the satellite subsystems from the extreme temperature conditions of the outer space.

The communications subsystems are the major elements of a communication satellite and the rest of the space craft is there solely to support it. Quite often it is only a small part of the mass and volume of the satellite. The communications subsystem consists of one or more antennas and communications receiver - transmitter units known as transponders. Transponders are of two types, Repeater or Bent pipe and processing or regenerative. In Repeater type, communications transponder receives the signals at microwave frequencies and amplifies the RF carrier after frequency conversion, whereas in processing type of transponder in addition to frequency translation and amplification, the RF carrier is demodulated to baseband and the signals are regenerated and modulated in the transponder. Analog communication systems are exclusively repeater type. Digital communication system may use either variety. The schematic diagrams of repeater type and regenerative type transponders respectively.

The actual reception and retransmission of the signals are however, accomplished by the antennas on board the satellite. The communications antennas on board the satellite maintain the link with the ground segment and the communications transponder. The size and shape of the communications antenna depend on the coverage requirements and the antenna system can be tailor made to meet the specific coverage requirements of the system.

Launch Vehicle

The function of the launch vehicle is to place the communication satellite in the desired orbit. The size and mass of the satellite to be launched is limited by the capability of the launch vehicle selected for launching the satellite. The satellite launch vehicle interface is also required to be provided as per the launch vehicle selected. Satellite launch vehicles are classified in two types i.e.

- Expendable
- Reusable

In expendable type the launch vehicle can be used only once and most of the launch vehicles are expendable type. Space Transportation System (STS) or Space Shuttle of NASA, USA is the only available operational reusable launch vehicle. Although most of the launches take place from ground, Sea Launch has embarked on the launching of satellites from off shore platforms and Peagasus launch vehicles can launch small satellites from aircrafts.

Launching of a satellite in orbit being a costly affair a number of programs have been undertaken by NASA to make the future launching of satellites in orbit as cost effective and routine as commercial air travel.

Satellite Control Centre

Satellite Control Centre performs the following function.

- Tracking of the satellite
- Receiving Telemetry data
- Determining Orbital parameters from Tracking and Ranging data
- Commanding the Satellite for station keeping
- Switching ON/OFF of different subsystems as per the operational requirements
- Thermal management of satellite.
- Eclipse management of satellite
- Communications subsystems configuration management.
- Satellite Bus subsystems configuration management etc.

GROUND SEGMENT

The ground segment of satellite communications system establishes the communications links with the satellite and the user. In large and medium systems the terrestrial microwave link interfaces with the user and the earth station. However, in the case of small systems, this interface is eliminated and the user interface can be located at the earth station. The earth station consists of:

- Transmit equipment.
- Receive equipment.
- Antenna system.

The schematic block diagram of an earth station.

In the earth station the base band signal received directly from users' premises or from terrestrial network are appropriately modulated and then transmitted at RF frequency to the satellite. The receiving earth station after demodulating the carrier transmits the base band signal to the user directly or through the terrestrial link.

The baseband signals received at the earth stations are mostly of the following types.

- Groups of voice band analog or digital signals
- Analog or digital video signals
- Single channel analog or digital signal
- Wide band digital signal.

In satellite communications, in early days FM modulation scheme was most frequently used for analog voice and video signal transmission. However, the trnd is now to use digital signals for both voice and video. Various digital modulation schemes like Phase Shift Keying (PSK) and Frequency Shift Keying

(FSK) are adopted for transmission of digital signals. The network operations and control centre for the communications network monitors the network operations by different users, distribution of different carriers within a transponder and allocation of bandwidth & EIRP of different carriers. Proper functioning of Network operations and control centre is essential where the number of users in the network is large. Network operations & control centre is also responsible for giving clearance to the ground system in respect of antenna radiation pattern, EIRP etc.

SATELLITE COMMUNICATIONS SERVICES

Different Satellite Communications services are classified as one way link and two way link. One way link from transmitter Tx to receiver Rx on earth's surface is shown in fig. below.

Examples of satellite services where the transfer of information takes place through one way link are:

- Broadcast Satellite Service (Radio, TV, Data broadcasting)
- Data Collection Service (Hydro meteorological data collection)
- Space operations service, (Tracking, Telemetry, Command)
- Safety services (Search & Rescue, Disaster Warning)
- Earth Exploration Satellite Service (Remote Sensing)
- Meteorological Satellite Service (Meteorological data dissemination)
- Radio Determination Satellite Service (Position location)
- Reporting Service (fleet monitoring)
- Standard frequency and time signal satellite service
- Space Research Service.

In two-way Satellite Communications link the exchange of information between two distant users takes place through a pair of transmit and receive earth stations and a satellite.

Examples of two-way satellite services are

- Fixed Satellite Service (Telephone, telex, fax, high bit rate data etc.)
- Mobile Satellite Service (Land mobile, Maritime, Aero-mobile, personal communications)
- Inter Satellite Service.
- Satellite News Gathering (Transportable and Portable)

A new class of two-way fixed satellite network service known as Very Small Aperture Terminal (VSAT) service has became very popular among business and closed users group communities. SAT networks are operated in two different configurations i.e. Mesh and Star. While in Mesh configuration a VSAT terminal can communicate with another VSAT terminal in a single hop connection, Star network involves two hops via satellite and the hub station.

6

Optus Satellite Business Services

This Service Description forms part of the Agreement under which Optus supplies the Service to you.

Rules of interpretation and capitalised terms which are used in this Service Description are defined either in the General Terms or in the attached Dictionary. Some important information about the Service:

Service Options	1 Satellite Service Options • SatWeb 1-Way • SatWeb 2-Way • SatData
Optus company supplying the Service	Optus Networks Pty Limited (ABN 92 008 570 330)

THE SERVICE

Optus will provide, and you must acquire, the Service in accordance with the Agreement for at least the Committed Term and any Holding Over Period. Optus will supply the Service with the Service Options you specify in the Application. Optus will continue to Supply the Service to you from the expiry of the Committed Term.

Either Party may Cancel

- The SatWeb 1 or 2 Way Private Services, and the SatData Private Service effective on or after the expiry of the Committed Term for the relevant Service by giving at least 90 days prior written notice; and
- Any other Service effective on or after the expiry of the Committed Term for that Service by giving at least 30 days prior written notice, to the other party.

Subject to clause 15 of the General Terms, if either party cancels the Service for any reason during the Committed Term, you must pay Optus the Cancellation Fee and any other outstanding charges.

SERVICE DESCRIPTION

The Service is:

- in the case of Satellite Service Options, a communications service provided via Satellite that allows users with VSATs situated within the Optus Satellite Footprint in Australia and New Zealand to:
  - Receive data via Satellite (SatWeb 1-Way); or
  - Receive and send data via Satellite (SatWeb 2-Way or SatData), where the data conforms to TCP/IP format and protocol conventions including Internet content, based on Service Feature selected; or
- in the case of the other specialised Satellite Service Options, a billing service that is used:
  - To rate your services and provide you with details of your usage; and
  - To rate your Customer Associates usage according to your tariffs and produce tax-invoices on your behalf (Bureau Billing).

SERVICE PROVISION: SERVICE FEATURES

You must select a Service Option in the Application. In relation to that Service Option, you must select at least one Service Feature in the Application. Optus will provide the Service to you based on those selections.

In the Application, you may select one or more of the following Service Features in respect of the SatWeb 1 and 2 Way services.

- SatWeb Internet Access: SatWeb Internet Access provides broadband access to the Internet. This feature utilises shared bandwidth on the Optus Satellite. User performance at any point in time is dependant on the number of other subscribers concurrently using the Service and the level of network traffic. This service feature does not include a public IP address.
- SatWeb Advance Service (also sold under the name "NetAccess Satellite"): The SatWeb Advance service provides broadband access to the Internet and also includes a public IP address as standard. This service feature utilises shared bandwidth on the Optus Satellite and is designed for customers wishing to extend their Intranets without the expense of dedicated bandwidth. User performance at any point in time is dependant on the number of other subscribers concurrently using the Service and the level of network traffic.
- SatWeb Private: SatWeb Private is for those customers requiring dedicated satellite bandwidth, which is scaled to meet their requirements in terms of the number of sites and data speeds. SatWeb Private is a satellite-carriage-only service and must be combined with a terrestrial connection from the BOC to some other location, or the Internet.

- SatStream Rapid File Distribution: SatStream Rapid File Distribution (RaFiD) is a shared service feature that enables the high speed delivery of electronic content to a community of interest. RaFiD can be used for database updates, software downloads, MPEG distribution, local store and forward services. Content is transmitted to your local PC.
  In the Application, you must select one of the following Service Features in respect of the SatData Service. Optus will provide the Service to you based on those selections.
- SatData Private: SatData Private provides you with dedicated bandwidth and is available at various data rates. It is a satellite-carriage-only service and must be combined with a terrestrial conne ction from the BOC to some other location. SatData Private does not include Internet access as part of its standard offering. For an additional charge, Optus will provide you with access to the Internet. The bandwidth allocated for Internet access is matched to the amount of Service bandwidth purchased by you. This Service does not include a content filtering service.
- SatData Private - Reservation: SatData Private – Reservation enables you to obtain dedicated bandwidth as and when required, subject to bandwidth availability. You may book the bandwidth by contacting the Optus reservation desk, on the number provided to you by Optus, one Business Day in advance of the booking (Scheduled Booking). At the time and for the duration of the Scheduled Booking, Optus enables you to access the bandwidth that you have booked.

SERVICE INSTALLATION

Optus will install the Purchased Equipment at the Location.

You acknowledge that:

- Optus may need to make physical modifications to the Location and open your Equipment in order to perform the installation; and
- The installation may invalidate any warranty in relation to your Equipment.

Any Loss you suffer arising from the installation is your risk and not Optus' risk and you will release Optus from all liability it may have to you in relation to such damage. Optus may charge you an additional charge for any non-standard installation or re-installation. You indemnify Optus against any Claim made against Optus by another person in connection with Optus' performance of the installation. If you plan to move to a new Location, you must provide Optus with at least 30 days prior written notice of your new address before you move. If the new Location is within the Optus Satellite Footprint, Optus may agree to install the Service at the new Location at an agreed time. Optus will charge you a Relocation Fee, and you must continue to pay the Monthly Charges for the Service even if the Service is temporarily unavailable because of your relocation. Optus is under no obligation to return the Location to its original condition (make good) after the Service is cancelled (for example, by removing wall sockets Optus has installed).

SERVICE CHARGES

You will be charged for the Services from the Service Start Date. The Charges for the Service will depend on:

- Tthe Service Option and Service Features selected by you in the Application; and
- Whether you modify the Service or cancel the Service at any time.

The Charges for the Service are fixed for at least the Committed Term. Optus may vary those Charges from time to time after the expiry of the Committed Term by giving you at least 30 days prior notice. If you exceed the amount of data allotted under your Service Plan, you must pay an Excess Data Charge.

INVOICING

Despite clause 5 of the General Terms, Optus will invoice you:

The Charges for the Service:

- Monthly in advance for all Monthly Charges; and
- Monthly in arrears, for any usage (Excess Data Charge) and/or One-Off Charges other than the Installation Charge which shall be dealt with pursuant of this Service Description.

On the Service Start Date:

- In advance for the first 2 months of Monthly Charges; and
- Any One Off Charge set out in the Application.

  You must pay all Charges for the Service within 30 days of the date of invoice. Each invoice issued or made available to you will detail:

  - Subject to clause 6.5 of the General Terms, the Charges payable less any Service Credits arising from the previous month; and
  - GST payable.

PURCHASED EQUIPMENT

You must:

- Purchase the Purchased Equipment from Optus for the price set out in

Application; and

- Pay the Installation Charge for that Purchased Equipment, on delivery or as agreed by the parties.

You are solely responsible for the operation, repair and Maintenance of the Purchased Equipment subject to any maintenance agreement with Optus or other Optus nominated third party. The Installation Charge depends on the Locations and includes the costs of cabling and labour.

The Installation Charge does not include:

- Installation at a Location that is not listed in the Application;
- Non-standard Installation (Optus reserves the right to charge a non-standard Installation Charge for non-standard installations);
- Any modification of your PC to make it comply with the specifications in this Service Description;
- Any operating system upgrade; or
- LAN software configuration.

YOUR EQUIPMENT

You are solely responsible for ensuring that your Equipment complies with the specifications set out in this Service Description. If your Equipment does not comply with these specifications you may not be able to successfully access, operate or use the Service and you may be refused assistance by the Help Desk. You must ensure that your Equipment does not cause Interference is installed and connected to the Purchased Equipment, only in the manner directed by Optus.

Optus may lease or arrange equipment to be leased to you (Leased Equipment) that would otherwise be Purchased Equipment. Any Leased Equipment remains the property of the relevant lessor pursuant to the terms and conditions of the applicable lease. You must at all times comply with the terms and conditions of the lease for the Leased Equipment.

OPERATIONAL ISSUES

You must provide, on request from Optus, details of the location of any Optus Equipment or Leased Equipment that is in your possession or control. If you request Optus to route traffic to a specific TCP/IP address, then Optus may, in its absolute discretion, agree to your request. In the case of private IP addresses (or private networks), Optus will discuss your requested address range with you to determine whether your request can be accommodated within the Optus Network. Optus may charge you an additional charge for this consultation.

Optus manages the allocation of all Services on all Satellites. You must comply with any specific operating conditions as advised by Optus from time to time. Compliance with these operating conditions may impact on your use of the Service. For operational reasons, you must advise Optus immediately of any variation in your use of the Service. You acknowledge that if Optus relocates a Service it may not be possible to duplicate all of the previous operating characteristics.

YOUR ACKNOWLEDGEMENTS AND OBLIGATIONS IN RELATION TO THE SERVICE

You acknowledge that the provision of the Service is subject to bandwidth availability. You may not re-supply the Service to any person, including Customer Associates without the written consent of Optus.

OBLIGATIONS IN RELATION TO REGULATIONS

The Parties must:

- Obtain and maintain in good standing all relevant Licences; and
- Ensure that all Customer Associates (in the case of you) and all Optus Associates (in the case of Optus) are at all relevant times duly Licensed.

OBLIGATIONS IN RELATION INTERNET ACCESS

In using the Service, you must:

- Comply with the Acceptable Use Policy;
- Ensure that the software you use in relation to the Service is properly licensed;
- Comply with any rules imposed by any third party whose content or services you access using the Service or whose network your data traverses; and
- Not infringe any person's intellectual property rights (such as by using, copying or distributing data or software without the permission of the owner).

You acknowledge that the Service relies for its operation on content or services supplied by third parties, who are not controlled or authorised by Optus.

Optus does not exercise any control over, authorise or make any warranty regarding:

- Your right or ability to use, access or transmit any content across the Internet (whether error-free, in time, or at all) using the Service;
- The availability, response times or other characteristics of the Service, except as set out in any service levels but subject to clause 16 of the General Terms;
- The accuracy or completeness of any content which you may use, access or transmit using the Service including, without limitation, any data which Optus may cache as part of the Service;
- The consequences of you using, accessing or transmitting any content using the Service, including without limitation any virus or other harmful software; or
- Any charges which a third party may impose on you in connection with your use of the Service.

You release and discharge Optus and each Supplier from any Loss, cost, liability or damage incurred or suffered by you which arises from or is connected with the matters referred to in this paragraph 12whether arising as a result of any act, omission or negligence of Optus, a Supplier or otherwise.

SYSTEM RECONFIGURATION

Optus may from time to time modify or vary the System including by:

- Modifying or varying the frequency of the Service;
- Relocating the Service to another part of the same Transponder;
- Relocating the Service to another Transponder (whether or not on the same Satellite and, in the case of a different Satellite, whether or not that Satellite is in the same orbit location as the original Satellite);
- Removing a Satellite from the System;
- Relocating a Satellite from one orbit location to another;
- Varying the orbit characteristics of a Satellite;
- Replacing a Satellite; or
- Introducing New Technology.

Optus will:

- Give you as much notice as is reasonably practicable concerning the timing, technical details of and reasons for any System Reconfiguration; and
- Provide to you all reasonable assistance you require to determine what adjustments or modifications to your Equipment are required as a result of a System Reconfiguration.

Optus excludes liability to you for, and you indemnify Optus against, any and all Loss suffered or incurred by you or any Customer Associates as a result of a System Reconfiguration, including Loss in connection with any replacement of or adjustments or modifications (including repointing) to your Equipment or Consumer Equipment, regardless of the cause of the Loss (including where caused or contributed to by any one or more negligent acts or omissions of Optus or any Optus Associate).

SERVICE RESTORATION AND SERVICE CREDITS

The procedures and principles that apply to restoration and discontinuance of Optus Satellite Services where one or more Optus Satellite Services fails to meet its or their Key Performance Indicators. For these purposes, the System is regarded as a single system of Transponders carried on several Satellites at nominated orbit locations, it being recognised that some Optus Satellite Services have requirements that can only be met using the particular characteristics of certain Transponders.

Each Optus Satellite Service:

- Is designated as either a Level 1 Service or a Level 2 Service; and

- Has a Priority Date,
- Which together will be used in accordance to determine how Optus Satellite Services which fail to meet their Key Performance Indicators will be restored.

If two or more Level 1 Services or Level 2 Services have the same Priority Date, Optus will determine (in its absolute discretion) the priority as between those Optus Satellite Services. If one or more Optus Satellite Service fails to meet its or their Key Performance Indicators (each an Affected Service) other than during any period in which Maintenance is being conducted then:

- Optus must analyse and attempt to determine the cause of the failure;
- Once Optus has determined the cause of the failure, Optus must use reasonable endeavours to restore that Affected Service and (if Optus is unable to do so through the use of redundant equipment) in accordance with the following principles:
  - Optus Satellite Services which can be restored will be restored on Transponders which Optus determines have suitable characteristics;
  - where practicable, Optus will aim to restore an Affected Service on a Transponder:
- At the same orbit location as the Transponder on which the Affected Service was previously provided; and
- With suitable characteristics as determined by Optus, and Transponder capacity will only be considered to be available for restoration purposes if Optus determines that it has suitable characteristics (and in determining whether a Transponder has suitable characteristics, Optus may have regard to its contractual obligations relating to Optus Satellite Services which are or may be carried on that Transponder);
  - Level 1 Services which are Affected Services (Affected Level 1 Services) will be restored in preference to Level 2 Services which are Affected Services (Affected Level 2 Services) and Occasional Services;
  - No operational Level 1 Service will be discontinued to restore any Affected Service;
  - Affected Level 1 Services will be restored through the use of any spare Transponder capacity until that capacity is exhausted and then through the direct or indirect displacement of Level 2 Services based on reverse Priority Date Order;
  - If there is insufficient Transponder capacity to restore all Affected Level 1 Services in accordance, Affected Level 1 Services will be discontinued based on reverse Priority Date Order;

- No operational Level 2 Service will be discontinued to restore any Affected Level 2 Service (except where that is the ultimate result of the restoration of an Affected Level 1 Service);
- If any Transponder capacity remains available after the restoration of Affected Level 1 Services as contemplated, Affected Level 2 Services will be restored through the use of any spare Transponder capacity until that capacity is exhausted;
- If there is insufficient Transponder capacity to restore all Affected Level 2 Services in accordance, Affected Level 2 Services will be discontinued based on reverse Priority Date Order;
- Where uplink services are provided in connection with Affected Services, the same restoration principles will apply to those uplink services; and
- Optus may discontinue any Occasional Service if necessary to restore any Affected Level 1 or Level 2 Service, Optus may treat capacity occupied by an Occasional Service as spare Transponder capacity.

Optus will endeavour to minimise disruption to Optus Satellite Services during any restoration activity contemplated, but you acknowledge that:

- Intermediate steps may need to be taken by Optus in the course of implementing the principles referred;
- Temporary interruption to the Service may arise;
- Restoration action will be taken in the order which Optus determines is the most operationally efficient;
- The principles reflect outcomes and do not require Optus to take restoration action in any particular order;
- Optus may, without penalty, discontinue the Service in the course of following the principles described, subject to any Service Credits that may be due in accordance; and
- Optus may need:
  - To modify or vary the frequency of one or more Optus Satellite Service;
  - To relocate one or more Optus Satellite Services to another part of the same Transponder;
  - To relocate one or more Optus Satellite Services to another Transponder (whether or not on the same Satellite and, in the case of a different Satellite, whether or not that Satellite is in the same orbit location as the original Satellite);
  - To remove a Satellite from the System;
  - To relocate a Satellite from one orbit location to another;
  - To vary the orbit characteristics of a Satellite;

   - To replace a Satellite; or
   - To introduce New Technology.

Despite anything:

- Optus will not be required to use or acquire additional Transponder capacity on a satellite other than a Satellite in order to restore any Affected Service; and
- Optus will not be obliged to restore any Optus Satellite Service where failure to meet Key Performance Indicators is a result of failure or incompatibility of your Equipment.

If the Service fails to meet its Key Performance Indicators, you must comply with any reasonable direction given by Optus which may remedy that failure. If the Service fails to meet its Key Performance Indicators, then unless that failure is due to an Excluded Event:

- Optus must give you the Service Credit for that failure; and
- If that failure continues for five days or more:
  - your obligation to pay the charges for the Service will be suspended until Optus restores or discontinues the Service in accordance; and
  - Optus will not be liable to give you a Service Credit while the obligation to pay the charges for the Service is suspended.

Service Credits:

- Will be in the form of a non-transferable deduction against the GST exclusive amount of Optus' next invoice for Monthly Charges; and
- Will, subject to paragraph 14.10, be your sole remedy for a failure by Optus to meet a Key Performance Indicator.

If:

- The Service fails to meet a Key Performance Indicator:
  - continuously for 21 days or more; or
  - on two or more separate occasions during any twelve consecutive months for an aggregate duration of 21 days or more; or
- The Service is discontinued pursuant for more than 21 days, either party may cancel the Service by giving notice to the other party.

Key Performance Indicators and Service Credits

If the Service fails to meet a Key Performance Indicator, Optus will give you a Service Credit. The Key Performance Indicator for the Service is an equivalent isotropic radiated power of 42.9dBW. All measurements are taken at the BOC under clear sky conditions.

Where the Key Performance Indicator is not met, Optus will give you a Service Credit in accordance with the following table:

Unavailable Time (minutes)	Equivalent Availability	Service Credit
0-1051	>= 99.80%	No Service Credit
1052 - 2102	>99.60% <99.80	One day's Monthly Charges for the Service
2102 - 3153	>99.40% <=99.60%	Two days' Monthly Charges for the Service
Greater than 3153	<= 99.40%	Three days' Monthly Charges for the Service

Unavailable Time means the period of time in minutes in each 12 month period from the Service Start Date that the Service or an Individual Service fails to meet a Key Performance Indicator less any period of time that the Service or that Individual Service (as the case may be) fails to meet its Key Performance Indicators due to:

- Scheduled Maintenance;
- System Reconfiguration; and
- Excluded Events.

The Service Credit will be applied against your next invoice. Service Credits are not redeemable for cash and in any month are capped at the monthly Service Charge.

EXCLUSIONS FROM THE SERVICES

The Services are not for resale. The Service does not include (and Optus has no obligation to provide):

- A VPN (unless otherwise agreed between the parties);
- A domain name server service and registration functions;
- Public IP addresses for you or Customer Associates unless explicitly included as part of the Service;
- Maintenance services (including, without limitation, for your Equipment);
- The dial up path from you to your ISP and from the ISP to the BOC;
- Gateway PC and associated applications to serve a LAN/WAN;
- Secure integration;
- E-mail services;
- The ISP service; or
- If you opt to connect your LAN to the Service, delivery of the Service beyond the Service Delivery Point.

Optus may in its absolute discretion offer to provide none, some or all of the Services, registration functions or addresses to you from time to time on terms to be agreed. You may wish to enter into a separate maintenance services agreement with Optus or pay for maintenance services on a call out basis. Any new services are subject to availability of Satellite capacity and Earth station infrastructure.

Help Desk

Optus will provide you with Help Desk services. The Help Desk is the single point of contact for support of the Service including billing enquiries and fault rectification. The Help Desk can be contacted on a telephone number provided by Optus from time to time. You must ensure that any person contacting the Help Desk is familiar with the technical configuration of your network. Before you contact the Help Desk, you must perform sufficient internal fault diagnostics to ensure that the problem is on the Satellite Component.

If:

- You contact the Help Desk regarding a fault, and when investigated Optus determines that the fault is not with the Satellite Component; and;
- Optus determines that if you had undertaken internal fault diagnostics then you would have ascertained that the fault was not with the Satellite Component,

then Optus will charge you for the services of the Help Desk at Optus' standard rates and charges from time to time (the current charge is $5 per call).

Indemnities

You indemnify Optus and each Optus Associate (those indemnified) from and against any and all Loss suffered or incurred by any of those indemnified in connection with Third Party Claims, regardless of cause (including where caused or contributed to by one or more negligent acts or omissions of those indemnified) Subject to clause 16 of the General Terms, Optus indemnifies you and your Related Corporation (those indemnified) against any and all Loss suffered or incurred by any of those indemnified in connection with any injury to or death of any Optus Personnel or any Personnel of an Optus Associate regardless of cause (including where caused or contributed to by any one of more negligent acts or omissions of any of those indemnified). Nothing in this clause excludes or limits the effect of any clause in the General Term.

Insurance

Because of the limited availability of insurance in respect of satellites and satellite based businesses, Optus may suffer detriment as a result of you or your Customer Associates seeking such insurance. You will not take steps to obtain insurance in respect of Consequential Loss or business interruption arising in connection with loss, failure or non performance of a Satellite Component without the prior written consent of Optus (which must not be unreasonably withheld). You must also ensure that your Customer Associates comply with this clause.

Dictionary

Acceptable Use Policy means the most current version of the Internet Acceptable Use Policy prescribing the rules and guidelines which you must comply with in using the Service available on the Optus website. BOC is the Broadcast Operations Centre, Belrose, NSW. Cancellation Fee will equal the present value of the product of: Minimum Net Activations, AvgServiceChg, Committed term and number of months remaining under the terms of the agreement, as per the following formula:

Cancellation Fee = PV [MNA * S * CT * M]

where

PV = Present value calculated using the Interest Rate defined below.

MNA = Minimum Net Activations, or set to value = 1 if not specified;

S = the average monthly service charge;

CT = Committed Term, or set to value = 12 if not specified.

M = Months remaining under the agreement.

The Interest Rate will be taken as the 'ANZ Reference Interest Rate' as published each Monday (or the next Business Day if the Monday is a public holiday) in the Australian Financial Review. If the rate is not quoted in the Australian Financial Review on any one or more of such days, there will be substituted the rate that any manager of Australian and New Zealand Banking Group Limited (acting as an expert and not as an arbitrator) determines on Optus' request as being the closest equivalent to that yield rate on the relevant day. Charges means the charges, set out in the Application, payable by you to Optus for the performance of the Services under this Agreement. Claim means any claim, demand, action or proceeding, whether in contract or tort, pursuant to statute, or otherwise which is made or brought in connection with:

- The provision or use by any person of the Services; or
- The Agreement.

Consumer Equipment means any signal reception or other equipment owned or controlled by end users of services provided by you.

Customer Associates means:

- Third parties (other than Optus or Optus Associates):
  - with whom you share, capacity of all or part of the Service, including End Users; or
  - to whom you provide goods or services using all or any part of the Service; and
- your Personnel; and
- the Personnel of those third parties.

Customer Associate Claim means any Claim made or brought by a Customer Associate or any person claiming by or through you. Customer Self Care means the application referred to in paragraph 1 of the VSAT Billing Services Option.

Equipment Areas means:

- In the case of Optus, any area within premises occupied or controlled by Optus where your Equipment is located from time to time; and

- In the case of you, any area at the Location occupied or controlled by you where Optus Equipment is located from time to time.

Excess Data Charge means the charge for [insert detail] as set out in the Application. Excluded Event in this Service Description also includes: an Intervening Event (other than a strike, lockout, labour dispute, work stoppage, embargo or other labour difficulty which is confined to Optus staff); your Conduct;

- Scheduled Maintenance, which does not exceed any period determined or agreed in accordance with Optus' standard practices and procedures from time to time;
- A failure of any equipment or service operated or provided by a person other than Optus or any Optus Associate; and
- A failure of the Service as a result of a Sun Transit.

General Terms means the most current version (from time to time) of the Optus General Terms available on the Optus Website at www.optus.com.au/standardagreements or on request. Help Desk means the help desk described in paragraph 17 of this Services Description. Holding Over Period for the Service means the period from expiry of the Committed Term until cancellation of the Service. Inroute means the path from your Equipment to the BOC. Installation Charge means the charge, set out in the Application, payable by you to Optus for the installation of the Equipment at the Location.

Interference means electromagnetic radiation emanating from any source (including the emission of signals from a Satellite or any other space satellite, the emission of spurious signals from defective equipment, unauthorised transmissions to a Satellite and the interaction of two or more signals authorised for transmission to a Satellite) which has or is capable of having a detrimental effect on a Satellite Component or equipment of Optus or a third party.

IP means Internet protocol.

ISP means Internet service provider.

Key Performance Indicators means:

- for the Service, excluding VBS-BB, the key performance indicators specified in paragraph 15.2;
- for VBS - Bureau Billing the key performance indicators specified in paragraph 9 of the VBS - BUREAU BILLING Service Option; and
- for any other Satellite Service, the key performance indicators, service levels or other performance measures for that Optus Satellite Service agreed between Optus and the recipient of that Optus Satellite Service.

Level 1 Service means:

- a service for which the service priority level specified in the Service Option is 'level 1'; and

- any Optus Satellite Service for which the service priority level is 'level 1' or its equivalent.

Level 2 Service means:

- a service for which the service priority level specified in the Service Option is 'level 2'; and
- any Optus Satellite Service, which is not a Level 1 Service.

Licences means all licences, authorities, consents, approvals or permits which Optus or you (as the case may be) or an Optus Associate or a Customer Associate (as the case may be) is required to hold or to have obtained in order to lawfully transmit or receive Telecommunications Traffic. Location means the site including the premises to which Optus will provide the Service, as specified in the Application. Maintenance means tests, adjustments or modifications to or in relation to any Satellite Component as may be necessary or convenient from time to time to maintain the System in efficient working order. Month means a calendar month.

Monthly Charges means the monthly recurring charges listed as Monthly Charges in the Application, which may include items such as satellite bandwidth, Internet access and/or maintenance New Technology means any method of encoding, encrypting, modulating, transmitting or receiving telecommunications signals having application or potential application to the provision of one or more Services which is not at the relevant time used by Optus in relation to the provision of the Services. NMS means the network management system.

Occasional Service means an occasional or part–time video interchange service, the SatData Private – Reservation service and any other Service provided via satellite which is not specified as a Level 1 Service or a Level 2 Service. One Off Charges means the charges listed as One off charges in the Application. These items are charged as and when incurred and typically relate to activities such as Service establishment (eg installation, Purchased Equipment), or relocation.

Optus Associates means:

- Each Related Corporation of Optus; and
- The Personnel of Optus or any Related Corporation of Optus.

Optus Associate Claim means any Claim made or brought by an Optus Associate (or any person claiming by or through Optus). Optus Employee Claim means any Claim made or brought by any Personnel of Optus or of any Optus Associate. Optus Equipment means the equipment owned or leased by Optus and used by Optus (in whole or in part) to provide the Services to you. Optus Satellite Footprint means the geographical coverage of the Satellite upon which the Service is connected from time to time. Optus Satellite Service means the Service and any other service provided via satellite by Optus to a customer other than an Occasional Service. Outroute means the path from the BOC to your Equipment. PC means personal computer.

Priority Date:

- Of a Service means the date specified in respect of that Service in item 3 of the Service Option; and
- Of an Optus Satellite Service means the priority date of that service as specified in, or computed in accordance with, any applicable agreement.

Priority Date Order of Services means the chronological order of the Priority Dates. Purchased Equipment means the equipment listed in the relevant Service Option that Optus will sell to you for those Services. Relocation Fee means the fee that Optus will charge a customer for connecting the Service to a site other than the Location. Satellite means a space satellite designed and procured for operation and use by Optus from time to time and used for the provision of the Services.

Satellite Component means any one or more of, or any part or combination of, the following:

- A Satellite;
- A Transponder;
- Optus' telemetry, tracking and control system;
- Optus' Satellite uplink equipment; and
- Any other equipment or systems used by Optus in the transmission or management of signals to or via a Satellite.

Satellite Service means a commercial service provided by Optus via Satellite from time to time. Scheduled Maintenance means maintenance, modification or an upgrade of the Optus Network:

- Of which affected customers have been given prior notice; and
- Which will result in all or part of the Service being temporarily impaired or unavailable.

Service Credit means the Service Credits (referred to as service rebates in the General Terms) specified in paragraph 14.2 in respect of the Service. Service Features means the features available to you in respect of SatWeb, SatData and VBS - Bureau Billing as set out in paragraphs 3.2, 3.3 and 3.4 respectively. Service Plan means the details and charges pertaining to the Service Feature, which are set out in the Application. Service Options mean the outlines of the various optional Services set out below.

Sun Transit means any period during which electromagnetic radiation from the sun causes Interference or results in a failure of the Service to meet its Key Performance Indicators.

System means the system by which Optus provides the Service and includes the Satellite Component. System Reconfiguration means any modification or variation to the System undertaken by Optus pursuant to paragraph 13.1. TCP means Transmission Control Protocol

Third Party Claim means any Claim made or brought against Optus or an Optus Associate by a third party, including a Claim made or brought by an End User or Transponder means that part of a Satellite which is capable of receiving, amplifying, translating and re transmitting Telecommunications Traffic, other than such a part which is not owned or leased by Optus or another subsidiary of Singtel Optus Pty Limited (ACN 052 833 208).

URL means Unique Resource Locator, which is more commonly known as a "web-address". VBS means the VSAT Billing System Service set out in the SatWeb 1-Way and 2-Way Service Options. VBS - Bureau Billing means the VSAT Billing System - Bureau-Billing Service set out in the VSAT Bureau Billing Service Option.

VSAT means very small aperture terminal. your Conduct means any breach of this Agreement by you and any act or omission (including negligent and other tortious acts or omissions) of you or a Customer Associate in connection with the use of the Service or the System. your Equipment means any equipment or facilities that are owned or controlled by you or a Customer Associates on you side of the Service Delivery Point and used in connection with the provision of the Services, including any Purchased Equipment.

SATWEB 1-WAY SERVICE OPTION

This section provides an outline of the SatWeb 1-way Service Option. Your Application details what Service Features you have selected in respect of the service. Service Option Summary

Item		
1.	Type of Service:	The SatWeb 1-way service is a PC-based, receive-only VSAT satellite communications system, that allows users with VSATs situated within the Optus Satellite Footprint to receive data via satellite.
2.	Priority Date:	Service Start Date.
3.	Service Priority Level:	Level 2
4.	Service Parameters	
	Satellite Position:	152 Degrees East
	Uplink site:	BOC
	Downlink Beam:	Optus National Beam
	Receive Speed:	Refer to Application
	Site Locations:	Refer to Application

WHAT IS SATWEB 1-WAY SERVICE?

SatWeb 1-Way incorporates a range of IP-based, satellite VSAT broadcast communications solutions, which depending on the Service Features selected, allow customers throughout Australia and New Zealand within the Optus Satellite Footprint to: a)receive Internet content via the satellite antenna; and/ or connect the service to and access the Service through a LAN (Support and configuration of the LAN remains the responsibility of you). The Service provides the Outroute from the Optus satellite Earth station to a VSAT antenna at the Location via a Satellite. You must supply and are responsible for the performance of, the Inroute from your Location, which may be to your ISP (for Internet access), or the Optus Network for other data carriage services.

The Service Delivery Points for the Service are:

- the point at which the Optus Network connects to the Internet (for Internet services), or the connection point to the Optus Network (for data carriage services); and
- the air interface at the VSAT antenna at the Location.

The Service uses TCP/IP as specified in the Internet Protocol Version 4 Specification RFC 791. Optus may also use performance enhancement protocols as required. You view your billed and unbilled usage of the Service via VBS.

Your Requirements to Access the Service

To be able to receive the SatWeb 1 Way Service, you must:

- Purchase or supply the Purchased Equipment listed in paragraph 5 of this Service Option;
- Ensure that your Equipment serving as the point of connection to the Purchased Equipment is operational and available by the proposed Service Start Date;
- Ensure that your Equipment complies with the specifications in this Service Option;
- Establish and properly maintain any connection necessary for you to connect the Purchased Equipment to the Service Delivery Point;
- Provide the Inroute between you and the Service Delivery Point either by establishing a connection to an ISP (for Internet services), or the Optus Network (data carriage services);
- Ensure your PC has the following minimum specifications:
- Pentium-class processor;
- 64 MB RAM (or as appropriate for the PC operating system in use);
- 500 MB of available hard-drive space;
- CD-ROM drive; and
- Operating systems: Microsoft® Windows® 98, Windows® 2000 or Windows® Me;
- Ensure software detailed in paragraph 1.5 of this Service Option, as applicable, is correctly configured on your PC; and
- Provide Optus, as required, with all routing information reasonably necessary to provide the Service.

If you wish to connect the Service to a LAN you must purchase a satellite receiver/router as detailed in paragraph 5 of this Service Option.

Structure of the Service

The data flow for the SatWeb 1-Way Internet service depicted in Diagram

1 of this Service Option, is for illustrative purposes only. The other Service Features represent variations to this basic data flow. The data flow for the SatWeb 1-Way Internet service works in the following way:

- You dial into your local/current ISP and send a request to view a website;
- The request is passed to the ISP, which is used as a routing path to the BOC;
- The ISP routes the request to the BOC, where it is identified and recognised as a SatWeb 1-Way Service authorised request;
- The request is 'de-encapsulated' and the content is obtained by the BOC from the Internet via existing high speed terrestrial backbones connected to the BOC; and
- The content is sent via Satellite to you, bypassing the ISP.

The Optus Network is controlled by an NMS located at the BOC. The Internet, the ISP and connections to and from the Optus Network and Internet or the ISP do not form part of the Optus Network, and are not provided by Optus as part of the Service (unless provided by Optus to you under a separate agreement).

The architecture and components of the Service, whether accessed through a single PC or a LAN. In the event of an inconsistency between Diagram 1 and any other part of this Service Description, the other part of the Service Description will prevail.

Network Responsibility

Optus is not responsible for any Service difficulty external to the Service Delivery Points of this Service Option including, without limitation, a Service difficulty:

- Within the network of an interconnecting supplier;
- Within the network of an interconnecting ISP;
- Resulting from your Equipment or applications, including its LAN or WAN;
- Resulting from non-authorised use or modification of your Equipment or the software; or
- Resulting from non-authorised use or modification of the Service.

Purchased Equipment

The following Purchased Equipment is required for SatWeb 1-Way:

- Receive-only satellite antenna;
- Low Noise Block;
- acceleration client software;
- Optus approved satellite receiver;
- Optus approved satellite receiver-router (for LAN connections); and
- Associated cabling.

Service Option, the following Purchased Equipment is required for all Service Features other than SatWeb Internet, Advance or Private:

- Applicable client software; and
- Applicable server software.

VSAT Billing System (VBS)

VBS provides you with online access to view:

- Usage of the Service for the previous day; and
- Usage statements for the previous month.

You access VBS via the Internet at the address (URL) advised by Optus from time to time. Optus will provide you with a username and password to enable you to access your account details.

In accessing the VBS you:

- Must comply with any rules imposed by any third party whose services you utilise;
- Acknowledge that Optus is not responsible where the performance of third party services used by you, prevent you from accessing VBS;
- Are responsible for any charges which a third party may impose on you in connection with your accessing VBS.
- Acknowledge that Optus does not exercise any control over, authorise or make any warranty regarding the availability or response times of VBS except;
- Are responsible for the safe-keeping of any user-passwords provided by Optus;
- You or any person using the VBS on your behalf will not do anything to interfere with the proper operation of the VBS or attempt to access information relating to another Optus customer.
- You release Optus from any obligations regarding losses incurred by you through errors and omissions in your utilisation of VBS.

VBS Restoration Targets

The following targets have been defined for VBS, which apply during the hours of 8am to 6pm EST on Business Days:

- Response Time of less than 1 hour. Response Time being the time between you reporting a fault and a case number being assigned.
- Restoration Time of less than 4 Business Days. Restoration time being the time between you reporting a fault and normal service being restored.

SATWEB 2 -WAY SERVICE OPTIONS

This section provides an outline of the SatWeb 2-Way Service Option. Your Application details what Service Features you have selected in respect of this Service. Service Option

Item	
1. Type of Service:	The SatWeb 2-way service is a VSAT satellite communications system, that allows users with VSATs situated within the Optus Satellite footprint to receive and send data via satellite.
2. Priority Date:	Service Start Date
3. Service Priority Level:	Level 2
4. Service Parameters	
Satellite Position:	152 Degrees East
Uplink site:	BOC
Downlink Beam:	Optus National Beam
Receive Speed:	Refer to Application
Site Locations:	Refer to Application

WHAT IS SATWEB 2-WAY SERVICE?

The Service is a range of stand-alone VSAT asymmetrical satellite communications solutions, which depending on the Service Features selected, allow customers throughout Australia and New Zealand within the Optus Satellite Footprint to:

- Access the Internet and send and receive data via a satellite antenna; and
- Connect the Service to and access the Service through a LAN (support and configuration of the LAN remains the sole responsibility of you).

The Service is carried to and from the satellite antenna via a two-way satellite connection, which means that you do not need a telephone line or connection to an ISP to receive the Service. The Service enables:

- Downloads using a shared broadband digital video broadcast-s (DVB-S) Outroute path; and
- Multiple shared frequency/ time division multiple access (FTDMA) inbound channels for the request path.

The Service uses TCP/IP as specified in the Internet Protocol Version 4 Specification RFC 791. Optus may also use performance enhancement protocols as required.

The Service Delivery Points for the Service are:

- the air interface at the VSAT antenna at the Location; and
- the point at which the Optus Network connects to the Internet (for Internet services), or leaves the Optus Network, including a point of carrier interconnect (for data carriage services).

You view your billed and unbilled usage of the Service via VBS.

Your Requirements to Access the Service

To be able to receive the Service, you must:

- Purchase or supply the Purchased Equipment listed in paragraph 6 of this Service Option;

- Ensure that your Equipment serving as the point of access to the Purchased Equipment is operational and available by the proposed Service Start Date;
- Ensure that your Equipment complies with the specifications in this Service Option;
- Establish and properly maintain any connection necessary for you to connect the Purchased Equipment to the Service Delivery Point;
- Ensure your PC has the following minimum specifications:
  - Pentium-class processor;
  - 64 MB RAM (or as appropriate for the PC operating system in use);
  - 500 MB of available hard-drive space;
  - CD-ROM drive; and
  - Operating systems: Microsoft® Windows® 98, Windows® 2000 or Windows® Me;
- provide Optus, as required, with all routing information reasonably necessary to provide the Service; and
- ensure that any software required for the Service is properly installed and configured on your PC.

Structure of the Service

The data flow for the SatWeb 2-Way Internet service is for illustrative purposes only. The other Service Features represent variations to this basic data flow. The SatWeb 2-Way Service consists of two main parts, the outbound and the in bound chains:

- The outbound chain manages the Outroute, from the BOC to your satellite antenna(s). It accepts the data transmissions (multi streams or unicast packets/segments) from various servers at the BOC or from the Internet itself. The composite stream is encapsulated into MPEG-2 frames using the MPE standard, and then up-linked to the Optus managed satellite as a DVB-S shared carrier for distribution to the remote sites; and
- The inbound chain manages the Inroute, from your satellite antenna(s) to the BOC via satellite. It accepts transmitted data from the satellite antenna. The received data is processed in the central processing unit of the inbound chain. The inbound chain also sends acknowledgments to the satellite antenna for the data received and timing information.

The entire system, including all components of the hub, as well as all remote sites, are controlled by a NMS located at the BOC. The architecture and components of this Service, whether accessed through a single PC or a LAN are depicted in Diagram 1 of this Service Option. In the event of inconsistency between Diagram and any other part of this Service Description, then the other part of this Service Description will prevail.

Sat Web Advance Service Notes

Optus does not guarantee the throughput of Internet services. The SatWeb Advance Service uses shared bandwidth on the Optus Satellite (unless otherwise agreed with Optus). User performance at any point in time is a function of the number of other subscribers concurrently using this Service and the level of network traffic. The SatWeb Advance Service is not designed for customers wishing to host web servers at the Location and their attention is drawn to the Acceptable Use Policy. It is recommended that customers with web servers obtain a service such as Optus Hosting, which would provide high-speed access for their customers and can be maintained remotely. The SatWeb Advance Service is supplied with a satellite acceleration client, which is a piece of software installed in your Equipment. This acceleration client manages the data flow over the Satellite improving the throughput. If you encrypt your data using, for example, IP Sec, the satellite acceleration will not be able to accelerate your traffic and your user performance will be affected. This Service does not support communication from one VSAT to another VSAT.

Network Responsibility

Optus is not responsible for any Service difficulty external to the Optus Network of this Service Option including, without limitation, a service difficulty:

- Within the network of an interconnecting supplier;
- Resulting from Equipment or applications, including its LAN or WAN;
- Resulting from non-authorised use or modification of your Equipment or the software; or
- Resulting from non-authorised use or modification of the Service.

Purchased Equipment

The following Purchased Equipment is required for SatWeb 2-Way Internet Access:

- Gilat Satellite Modem and associated drivers;
- Gilat Solid State Power Amplifier SSPA;
- Gilat Low Noise Block;
- 2-way VSAT antenna;
- Acceleration client software (as required)
- associated cabling

Service Option, the following Purchased Equipment is required for all Service Features other than SatWeb Internet, Advance or Private:

- Applicable client software; and
- Applicable server software.

In accessing the VBS you:

- must comply with any rules imposed by any third party whose services you utilise;

- Acknowledge that Optus is not responsible where the performance of third party services used by you, prevent you from accessing VBS;
- Are responsible for any charges which a third party may impose on you in connection with your accessing VBS.
- Acknowledge that Optus does not exercise any control over, authorise or make any warranty regarding the availability or response times of VBS except;
- Are responsible for the safe-keeping of any user-passwords provided by Optus;
- You or any person using the VBS on your behalf will not do anything to interfere with the proper operation of the VBS or attempt to access information relating to another Optus customer.
- You release Optus from any obligations regarding losses incurred by you through errors and omissions in your utilisation of VBS.

VBS Restoration Targets

The following targets have been defined for VBS, which apply during the hours of 8am to 6pm EST on Business Days:

- Response Time of less than 1 hour. Response Time being the time between you reporting a fault and a case number being assigned.
- Restoration Time of less than 4 Business Days. Restoration time being the time between you reporting a fault and normal service being restored.

SATDATA SERVICE OPTION

An outline of the SatData Service Option. Your Application details what Service Features are applicable to your Service. Service Option Summary

Item	
1. Type of Service:	The SatData service is a VSAT satellite communications system, that allows users with VSATs situated within the Optus Satellite footprint to receive and send data via satellite.
2. Priority Date:	Service Start Date
3. Service Priority Level:	Level 2
4. Service Parameters	
Satellite Position:	160 degrees East
Uplink site:	BOC.
Downlink Beam:	Optus National Beam
Receive Speed:	Refer to Application
Site Locations:	Refer to Application
5. Special Conditions:	Set out in Annexure A to this Service Option in respect of the SatData Private Reservation service

WHAT IS SATDATA SERVICE?

The SatData Service is a range of stand-alone VSAT asymmetrical satellite communications solutions, which depending on the Service Features selected, allows customers throughout Australia and New Zealand within the Optus Satellite Footprint to:

- Connect the Service to and access the Service through a LAN (support and configuration of the LAN remains the sole responsibility of you) and/or;
- Access the Internet.

The Service Delivery Points for the Service are:

- The air interface at the VSAT antenna located at the Location; and
- The point at which the Optus Network connects to the Internet (for Internet services), or leaves the Optus Network, including a point of carrier interconnect (for data carriage services).

Depending on the Service Feature chosen, Customers utilise the SatData service to access the Internet or when combined with a separately purchased Optus terrestrial service (eg Frame, ATM) have connectivity to their data centre.

Your Requirements to Access the Service

To be able to access the Service, you must:

- purchase or supply the Purchased Equipment of this Service Option;
- ensure that your Equipment serving as the point of access to the Purchased Equipment has an Ethernet (IEEE 802.3) interface;
- ensure that your Equipment is operational and available by the proposed Service Start Date;
- establish and properly maintain any connection necessary for you to connect the Purchased Equipment to the Service Delivery Point;
- ensure that any PC accessing the Service has the following minimum specifications
  - Pentium-class processor;
  - 64 MB RAM (or as appropriate for the PC operating system in use);
  - 500 MB of available hard-drive space;
  - CD-ROM drive;
  - Operating systems: Microsoft® Windows® 98, Windows® 2000 or Windows® Me; and
- provide Optus as required, with all routing information reasonably necessary to provide the Service.

Structure of the Service

The data flow for the SatData service depicted in Diagram 1 of this Service

Option, is for illustrative purposes only. The other Service Features represent variations to this basic data flow. The Service is based on the Shiron InterSKY satellite platform, which consists of a hub at the Optus earth station and remote VSAT sites. It uses TCP/IP as specified in the IP Version 4 Specification RFC 791. Optus may also use performance enhancement protocols as required.

The Service provides downloads using a shared broadband DVB-S Outbound and uploads via multiple FDMA/DAMA Inbound channels. Service delivery consists of two main parts: the outbound and the Inbound chains. The Outbound chain, from the hub to the remote site, manages the Outroute. It combines the data transmissions from various services at the hub, or from the Internet into a composite data stream. This composite stream is then encapsulated into MPEG-2 frames using the MPE standard, and then uplinked to the Optus Satellite for distribution to Customer Locations The Inbound chain, from the remote site to the hub, manages the Inroute. It comprises your satellite antenna transmitting data to the Optus satellite and back to the SatData hub. The received data is processed in the central processing unit of the inbound chain and distributed as appropriate. This chain also incorporates acknowledgements set to your's satellite antenna for data received and timing information. The entire system, including all components of the hub, as well as all remote sites are controlled by the NMS located at the BOC. The architecture and components of the Service of this Service Option. In the event of inconsistency between Diagram 1 of this Service Option and any part of this Service Description, the other part of this Service Description will prevail.

NETWORK RESPONSIBILITY

Optus is not responsible for any Service difficulty external to the Optus Network, without limitation, a Service difficulty:

- Within the network of an interconnecting supplier;
- Within your Equipment or applications, including its LAN or WAN;
- Resulting from non-authorised use or modification of your Equipment or the Software; or
- Resulting from non-authorised use or modification of the Service.

Purchased Equipment

The following Purchased Equipment is required for each Service Feature available under this Service Option:

- Shiron InterSkyTM Satellite Modem and associated drivers;
- Shiron Solid State Power Amplifier SSPA;
- Shiron Low Noise Block;
- 2-way VSAT antenna;
- Acceleration client software (as required); and
- Associated cabling

7

Managed Video Conferencing Service and Equipment

MANAGED VIDEO CONFERENCING SERVICE

WHAT IS THE MANAGED VIDEO CONFERENCING SERVICE?

The Managed Video Conferencing Service consists of one or more of the following service components, depending on what you choose:

- Equipment, which you can:
- Purchase from us;
- Rent from us;
- Apply to lease from us; or
- Acquire separately, provided that the equipment has been approved by us for use with your Managed Video Conferencing Service;
- Planning, design, installation and commissioning;
- Wall mount deployment;
- Facilities management;
- Training;
- Software;
- Managed services; and
- Support and maintenance.
- Access service

You must separately acquire from us the access service necessary for the connection and carriage of video conference calls. We cannot provide the Managed Video Conferencing Service to you if you do not have an access service between your sites.

It is your responsibility to choose and maintain your access service separately. The terms (including fees and charges) for your access service are separate from and in addition to the terms (including fees and charges) for your Managed Video Conferencing Service.

USING YOUR MANAGED VIDEO CONFERENCING SERVICE

Eligibility

The Managed Video Conferencing Service is not available to Telstra Wholesale customers or for resale.

Fees and charges

The fees and charges for your Managed Video Conferencing Service are set out in your application form or separate agreement with us.

Minimum Term

You must take the Managed Video Conferencing Service for a minimum term of 12 months.

Cancelling your Managed Video Conferencing Service

If either your Managed Video Conferencing Service or your access service is cancelled (for any reason), the other service is not cancelled automatically. You have to cancel it yourself separately. If your Managed Video Conferencing Service is cancelled before the end of your minimum term, we may charge you an amount equal to:

- 50% of the monthly services charge applicable to your chosen service components, multiplied by the number of months remaining in the minimum term (this does not apply where we cancel your Managed Video Conferencing Service when you are not in breach or where you cancel your Managed Video Conferencing Service because we are in breach);
- The remaining balance of the price of any purchased equipment;
- If you rented equipment from us as part of your Managed Video Conferencing Service, 100% of the rental monthly charges, multiplied by the number of months remaining in the minimum term (this does not apply where we cancel your Managed Video Conferencing Service when you are not in breach or where you cancel your Managed Video Conferencing Service because we are in breach); and
- If you leased equipment from us as part of your Managed Video Conferencing Service, 100% of the monthly lease charges, multiplied by the number of months remaining in the minimum term (this does not apply where we cancel your Managed Video Conferencing Service when you are not in breach or where you cancel your Managed Video Conferencing Service because we are in breach).

You acknowledge that this amount is a genuine pre-estimate of the loss we are likely to suffer.

Email Alerts

From time to time, we may provide you with email alerts regarding Software Updates and New Releases, security issues and general product information. We will use the email address that you provide us. You consent to receiving these emails.

Third Party Suppliers

You acknowledge that we may purchase some components of your Managed Video Conferencing Service from third party suppliers. If one of our third party suppliers suspends, cancels or terminates a service that we rely on to provide you with your Managed Video Conferencing Service, we may:

- Replace or modify your Managed Video Conferencing Service; or
- Suspend, cancel or terminate your Managed Video Conferencing Service or the affected part.

We will give you as much notice as is reasonably possible in the circumstances.

Monitoring, Recording and Streaming of Video Conferences

Depending on your chosen equipment, your users may be able to record and stream video conferences. You acknowledge that we (and our subcontractors) and your users may monitor your video conference (including connection status), record meetings and collect and use identifying information about the participants using the service, such as a name or document that is displayed, transmitted, processed, or stored as part of a meeting or meeting record.

Certain laws require individuals to give their prior consent to the recording of their communications and/or restrict the collection, storage and use of the information that identifies them. You agree to:

- Comply with all applicable laws;
- Provide any reasonable assistance we request to assist us in complying with our obligations under applicable laws; and
- Obtain all necessary consents and make all necessary disclosures before you or any user accesses or uses the recording or streaming capability.

EQUIPMENT

You can:

- Purchase equipment from us;
- Rent equipment from us;
- Apply to lease equipment from us; or
- Use your own equipment, provided that the equipment has been

approved by us for use with the Managed Video Conferencing Service.

Purchase of Equipment

If you purchase equipment from us for use with your Managed Video Conferencing Service, the following terms apply.

You may purchase equipment from us either by:

- Paying us the upfront purchase price; or
- If we agree, pay us a monthly charge over a fixed term.

Title to the equipment will pass to you once you have paid us in full for the equipment. Until that time, you hold the equipment on our behalf and must promptly return the equipment if we ask you to. Risk in the equipment transfers to you on delivery. Until you have paid us in full for the equipment, you must treat the equipment as rental equipment and the obligations set out below in relation to rental equipment apply to you. In addition to any other rights we may have, if you do not pay us the relevant charges for equipment as set out in your agreement with us, then you must:

- Deliver the equipment, back to us, at your cost, in good working order and condition (reasonable wear and tear excepted) to such place in Australia as we may reasonably direct. If the equipment is being returned from outside Australia you must ensure that all applicable import and customs taxes, charges and duties are paid in full; and
- If applicable, immediately pay to us any applicable early termination charge or costs associated with restoring or replacing the equipment.

If you do not deliver the equipment as you are required, then:

- We may, or our agent may, enter any sites where we believe the equipment may be located for the purpose of recovering it; and
- You must pay us for any costs which we may reasonably incur in recovering or attempting to recover the equipment.

If you purchase equipment from us or use your own equipment, there is a risk that over time we may no longer support that equipment. We will notify you if this occurs. While this equipment may continue to function, we are unable to offer any guarantees as to the quality, performance or functioning of equipment which we no longer support.

Rental of Equipment

If you rent equipment from us, the following terms apply. Title to the equipment does not pass to you at any time. Risk in the equipment transfers to you on delivery. If you cancel an order for equipment after we have ordered it from our supplier, in addition to any other rights we may have, we may require you to pay us for the equipment that has been ordered. You must take reasonable care of the equipment and agree to pay us for any damage to the equipment that is caused or contributed to by you (including any of your employees, contractors or agents). If the equipment is damaged, destroyed, lost or stolen at any time, then we may charge you an additional fee to repair or replace the equipment. You must not modify or repair the equipment without our consent. If you make any modifications or repairs to the equipment and it impairs the condition of the equipment or diminishes its use or value, then we may charge you an additional repair fee. If you replace any part of the equipment, you must ensure that the replacement part is of equal or better quality than the removed part. If any part of the equipment is replaced or you add any new parts to the equipment, then that replacement or new part will become part of the equipment (and is our property). You must ensure that any replacement or new part is compatible with the equipment. You must only use the equipment, including any replacement equipment provided by us:

- In connection with your Managed Video Conferencing Service at your nominated sites;
- For the purpose for which it was designed;
- In a manner that is contemplated by the manufacturer and in accordance with the manufacturer's manuals and recommendations;
- In compliance with all relevant laws;
- In accordance with our reasonable directions;
- In a suitable environment for the correct operation of the equipment; and
- With a suitable network service and must not attach or enable connection with any other incompatible equipment or service.

You must:

- Maintain a suitable environment for the correct operation of the equipment, including the availability of necessary auxiliary services for the correct operation of the equipment;
- Ensure that the equipment is kept safe and secure to our reasonable satisfaction, including protecting the equipment from electrostatic interference and power surges;
- Ensure that the equipment is kept in good order and repair (if you do not, we may require you to reimburse us for the cost of restoring the equipment);
- Comply with our reasonable directions to protect our ownership of the equipment; and
- allow us (or our subcontractors) to inspect the equipment on reasonable notice.

You must not:

- Attempt to sell, dispose of or encumber the equipment in any way; or
- Remove, cover, alter or otherwise tamper with any labels or identifying markings on the equipment.

You must promptly notify us if any equipment is lost, stolen, damaged, destroyed or otherwise unfit or unavailable for use. If your Managed Video Conferencing Service is cancelled or terminated for any reason, then you must:

- Within 14 days of cancellation or termination, deliver the equipment, back to us, at your cost, in good working order and condition (reasonable wear and tear excepted) to such place in Australia as we may reasonably direct. If the equipment is being returned from outside Australia you must ensure that all applicable import and customs taxes, charges and duties are paid in full; and
- If applicable, immediately pay to us any early applicable early termination charge or costs associated with restoring or replacing the equipment.

If you do not deliver the equipment as you are required, then:

- We (as the owner of equipment) may, or our agent may, enter any sites where we believe the equipment may be located for the purpose of recovering it; and
- You must pay us for any costs which we may reasonably incur in recovering or attempting to recover the equipment.

There is a risk that over time we may no longer support your equipment. We will notify you if this occurs. On request we can recommend replacement equipment, but we cannot guarantee that this will be available at the same price as your existing equipment. While equipment may continue to function, we are unable to offer any guarantees as to the quality, performance or functioning of equipment which we no longer support. If you rent equipment from us, at the end of your minimum term you may apply to purchase the equipment from us and we will agree a price for the equipment with you. If you apply to purchase the equipment from us following clause 4.23 and we agree the price:

- Title to the equipment will pass to you once you have paid us in full for the equipment. Until that time, you hold the equipment on our behalf and must promptly return the equipment if we ask you to; and
- Until you have paid us in full for the equipment, the equipment will remain rental equipment and the obligations in relation to rental equipment (including payment of rental charges) will continue to apply to you.

Lease of Equipment

If you lease equipment from us, the following terms apply. Title in the equipment does not pass to you until all the lease charges have been paid in full. Risk in the equipment transfers to you on delivery. If you cancel an order for equipment after we have ordered it from our supplier, in addition to any other rights we may have, we may require you to pay us for the equipment that has been ordered. You must take the Managed Video Conferencing Service in respect of the equipment for a minimum term of 24 months.

You must acquire support and maintenance services from us in respect of the equipment for the same minimum term as the lease. If we accept your application for additional equipment a new minimum term will apply to that equipment. You must take reasonable care of the equipment and agree to pay us for any damage to the equipment that is caused or contributed to by you (including any of your employees, contractors or agents). If the lease equipment is damaged, destroyed, lost or stolen at any time, then we may charge you an additional fee to repair or replace the equipment. You must not modify or repair the equipment without our consent. If you make any modifications or repairs to the equipment and it impairs the condition of the equipment or diminishes its use or value, then we may charge you an additional repair fee. If you replace any part of the equipment, you must ensure that the replacement part is of equal or better quality than the removed part. If any part of the equipment is replaced or you add any new parts to the equipment, then that replacement or new part will become part of the equipment (and is our property). You must ensure that any replacement or new part is compatible with the equipment. You must only use the lease equipment, including any replacement equipment provided by us:

- In connection with your Managed Video Conferencing Service at your nominated sites;
- For the purpose for which it was designed;
- in a manner that is contemplated by the manufacturer and in accordance with the manufacturer's manuals and recommendations;
- In compliance with all relevant laws;
- In accordance with our reasonable directions;
- In a suitable environment for the correct operation of the equipment; and
- With a suitable network service and must not attach or enable connection with any other incompatible equipment or service.

You must:

- Maintain a suitable environment for the correct operation of the equipment, including the availability of necessary auxiliary services for the correct operation of the equipment;
- Ensure that the equipment is kept safe and secure to our reasonable satisfaction, including protecting the equipment from electrostatic interference and power surges;
- Ensure that the equipment is kept in good order and repair (if you do not, we may require you to reimburse us for the cost of restoring the equipment);
- Comply with our reasonable directions to protect ownership of the equipment; and

- Allow us (or our subcontractors) to inspect the equipment on reasonable notice.

You must not:

- Attempt to sell, dispose of or encumber the equipment in any way; or
- Remove, cover, alter or otherwise tamper with any labels or identifying markings on the equipment.

You must promptly notify us if any equipment is lost, stolen, damaged, destroyed or otherwise unfit or unavailable for use. If your Managed Video Conferencing Service is cancelled or terminated for any reason, then you must:

- Within 14 days of cancellation or termination, deliver the equipment, back to us, at your cost, in good working order and condition (reasonable wear and tear excepted) to such place in Australia as we may reasonably direct. If the equipment is being returned from outside Australia you must ensure that all applicable import and customs taxes, charges and duties are paid in full; and
- If applicable, immediately pay to us any early applicable early termination charge or costs associated with restoring or replacing the equipment.

If you do not deliver the equipment as you are required, then:

- We (as the owner of equipment) may, or our agent may, enter any sites where we believe the equipment may be located for the purpose of recovering it; and
- You must pay us for any costs which we may reasonably incur in recovering or attempting to recover the equipment.

There is a risk that over time we may no longer support your equipment. We will notify you if this occurs. On request we can recommend replacement lease equipment, but we cannot guarantee that this will be available at the same price as your existing equipment. While the lease equipment may continue to function, we are unable to offer any guarantees as to the quality, performance or functioning of equipment which we no longer support.

Your Equipment

For a list of equipment approved by us for use with the Managed Video Conferencing Service, please contact your Telstra representative.

Delivery of Equipment

We will use our best endeavours, but do not promise, to deliver the equipment to you on or before the date we agree to deliver or make the equipment available to you. Our standard hours for delivery of equipment are during business hours. If you request us to deliver and install equipment outside business hours, we may charge you an additional charge. If you cancel an order for equipment after we have ordered it from our supplier but before delivery, in addition to any other rights we may have, we may require you to pay us for the equipment that has been ordered.

PLANNING, DESIGN, INSTALLATION AND COMMISSIONING

Planning and Design

Before we install and commission your Managed Video Conferencing Service, we will work with you to prepare a high level Design Plan of your Managed Video Conferencing Service, including:

- A description of your telecommunications network (including any required changes to your sites and existing network);
- Recommended access methods;
- Hardware requirements;
- configuration, interface and system (including software) requirements;
- Conferencing equipment (including room layout and design);
- Site and connectivity specifications; and
- Any other recommendations.

On request, or where we determine that installation of your Managed Video Conferencing Service is 'complex', we will work with you to prepare a detailed Design Plan of your Managed Video Conferencing Service. We will notify you that that your Managed Video Conferencing Service is 'complex' if any one or more of the following are required:

- Any change to building works;
- Complicated audio visual integration;
- Deployment of video conferencing network customer premises equipment (for example, bridges or servers);
- Large scale deployments;
- Non-standard deployments; or if you acquired your Managed Video Conferencing Service on and from 8 July 2011:
- The services of any third party designers or installers; or
- Equipment which is designed and prefabricated overseas.

We will submit the Design Plan to you for your approval. Once you approve the Design Plan, any subsequent change may be considered a change of scope which could impact the overall design, charges and implementation timeframes of your Managed Video Conferencing Service.

You must provide us with all reasonable assistance and information we require to prepare the Design Plan, including:

- A detailed description of your sites;
- Any existing network diagrams and details of any existing communications infrastructure and equipment you have; and
- Provide a single point of contact for project co-ordination and resolution of on-site requirements.

You must ensure that all the information you provide us is accurate, complete and up to date, otherwise your Design Plan may be unsuitable and contain errors. We may charge you an additional charge for any additional work we are required to perform as a result of the information being inaccurate, incomplete or out-of-date. Unless otherwise expressly stated in the Design Plan, the charges and proposed timeframes in the Design Plan exclude:

- Any costs associated with delays resulting from the presence of asbestos or other hazardous materials and the removal of such material;
- Installation and activation of your power cabling and carriage service;
- Any costs associated with lost time due to re-scheduling, industrial disputes, or delays due to circumstances outside our reasonable control; and
- Cutting channels or core holes into concrete.

Site Survey

On request, or following our recommendation, we will perform a physical site survey of the sites to assess whether they (including your network, systems, software and equipment) are suitable for your Managed Video Conferencing Service. We may charge you an additional charge for this service, as set out in your separate agreement with us. If following the site survey we recommend any changes to your network, systems, software or equipment, you must make these changes before we will provide your Managed Video Conferencing Service. On request, we (or our subcontractors) may assist you with modifying the sites for video conferencing. We may charge you an additional charge for this service which will be advised on application. You must obtain (at your cost) all necessary consents in order for us to perform this service.

Your Sites

It is your responsibility to ensure that your sites are suitable for video conferencing (including providing sufficient space, suitable power, cabling, data cabling points and electrical outlets) and ensuring that you have a suitable wall structure to mount equipment. You must provide us (and our subcontractors) with sufficient access to your sites, video conference rooms, CPE accounts and passwords necessary to access your equipment and provide your Managed Video Conferencing Service. Additional charges may apply if we (or our subcontractors) attend your site but are unable to access your site or equipment because of a failure by you.

You must:

- Ensure that our personnel (and our subcontractors) are provided with a safe working environment when working on the sites, including sufficient working space and facilities;

- Provide us with all assistance and access to information, materials, your network and systems at the sites as reasonably requested by us; and
- Where applicable, obtain (at your cost) all third party consents necessary for us to access and use the sites and any materials requested by us.

Installation and Commissioning

We will install and commission your video conference service (including equipment) in accordance with the Design Plan. You agree to provide us with reasonable assistance during the installation and commissioning process.

Testing

Following installation and commissioning of your Managed Video Conferencing Service, we will test it to check that it operates correctly. We will notify you when we have completed testing. You must then notify us in writing whether you consider that your Managed Video Conferencing Service is operating correctly. We will only activate your Managed Video Conferencing Service once you notify us that it is operating correctly. If your Managed Video Conferencing Service does not work correctly, we will try to correct the problem, and we will retest it.. If your Managed Video Conferencing Service still does not work correctly, you may:

- Accept your Managed Video Conferencing Service as is, subject to the agreeing any changes to the charges to reflect any reduced level of functionality (or other deficiencies) in your Managed Video Conferencing Service; or
- Cancel your Managed Video Conferencing Service without the payment of any early termination charges.

WALL MOUNT DEPLOYMENT

If you apply for a wall mount deployment option, you:

- Must ensure that your sites have an appropriately solid and stable dual panel wall with studs no further apart than one every 60 cms, certified as load-bearing sufficient to support the weight of the screens (maximum weight of one screen at 52Kg, or two screens at 104Kg) and all related cabling, brackets and other associated equipment ("Suitable Wall Structure"); and
- Acknowledge and agree that you have selected a wall mount deployment option at your own risk and that to the full extent allowed by law, we do not guarantee, and are not liable in connection with, the wall mount deployment (including that the

relevant wall has sufficient load bearing for the required equipment and related materials).

Screens larger than 60 inches are not installed as standard. If you require larger screens, please contact your Telstra representative. Once we have accepted your application, we (or our subcontractor) will attend your site to begin the wall mount deployment ("Initial Site Visit"). If during the Initial Site Visit, we (or our subcontractor) reasonably determine that your site is not currently suitable for a wall mount deployment, we may contact you to arrange a subsequent site inspection ("Subsequent Site Inspection") but this does not limit your acknowledgement above.

During or after the Initial Site Visit or the Subsequent Site Inspection (as applicable), we (or our subcontractor) may provide you with:

- An additional wall-mount deployment fee in addition to the installation charges;
- Instructions for you to remedy the site for a wall mount deployment (including instructions to ensure a Suitable Wall Structure at the site); or
- A solution other than a wall mount deployment for your Managed Video Conferencing Service (such as a standard deployment).

If you accept the additional wall-mount deployment fee or choose to proceed with the remedial work, prior to any installation work commencing, you must:

- Confirm that authorisation to proceed with a wall mount has been secured from the owner of the property (if you are not the owner), and provide this to us prior to commencement of installation;
- On request, organise for us (or our subcontractor) to meet with the building owners to discuss the wall mount (where applicable); and
- Confirm to us that the site has been remedied in accordance with the instructions given above (where applicable).

If at any time you do not proceed with the wall mount deployment option at a particular site for any reason, or if our technician determines that a particular site is not suitable for the wall mount deployment option, you must:

- Pay the charge set out in your application form or separate agreement with us for each site visit (including the Initial Site Visit and the Subsequent Site Inspection); and
- Purchase the equivalent package with the standard deployment option (mounted on a stand). For example, if you had ordered an HDX7000 to be installed with 2 x 52 inch LCD screens (Wall Mounted), you agree to purchase the equivalent HDX7000 standard deployment package with 2 x 52 inch LCD screens.

You may apply to us for a quote to decommission the wall mount installation at the end of the contract period. Unless otherwise specified in your application form or separate agreement with us, decommissioning is not included.

FACILITIES MANAGEMENT

You may apply for Facility Management service for your Video Conference service either on an ongoing or a once off basis by completing and returning the application form to us at least 15 business days in advance of your preferred start date. If you apply for the Facilities Management service on less than 15 business days notice, we cannot guarantee that we will meet your request. If we accept your application, we will provide a video conferencing consultant at your nominated site(s) to provide assistance with your Video Conference service (such as scheduling, call initiation and problem solving, and liaising with the Telstra Helpdesk if required).

You must:

- Ensure that our video conferencing consultant is provided with a safe working environment when working at your sites, including sufficient working space and facilities;
- Provide our video conferencing consultant with all reasonable assistance and access to information, materials, your network and systems at your sites as requested by us; and
- Where applicable, obtain (at your cost) all third party consents necessary for our video conferencing consultant to access and use your sites and any materials requested by us.

TRAINING

On request, we will provide you with training in relation to operating and using your video conferencing equipment for an additional charge. The level of training (including the charges for the training) will be advised on application. If you acquired your Managed Video Conferencing Service on and from 8 July 2011 and you apply for training (other than basic training at the time of installation) which requires us to attend your site(s), we may charge you for any transport and accommodation costs incurred by our trainers in addition to the charges for the training. We may provide you with other training at no additional charge from time to time. We will include the charges for any training you request in your first bill after we have commissioned your Managed Video Conferencing Service, so you may be billed before we have delivered the training.

SOFTWARE

We may provide you with Software that you will need to download onto yourequipment. If we do this, we will either:

- Grant you a non-exclusive licence to use the Software for the purpose of using your Managed Video Conferencing Service; or
- Procure the right for you to use the Software, in which case the Software will be provided on the terms notified to you in your separate agreement with us.

If we provide you with Software, we will notify you of any Updates and New Releases to the Software when they become publicly available. We may provide you with Updates and New Releases within a reasonable period of time after we have tested them for compatibility. We may provide Updates at no charge, but may charge you an additional charge for any New Releases we provide. You can ask us questions about the installation of the Software via the helpdesk.

If you refuse to accept an Update or New Release as recommended by us, you acknowledge that we may not be able to provide all aspects of your Managed Video Conferencing Service, or maintain and support your Managed Video Conferencing Service.

MANAGED SERVICES

Asset Management

On request, we will maintain (updated with version control) a database to track the:

- Serial number;
- Make and model;
- Description;
- Date of installation; and
- Location installed,
- Of all equipment managed by us as part of your Managed Video Conferencing Service ("Asset Register").

You may access the information contained in the Asset Register via a web portal (where available), otherwise we will provide the information to you in a suitable text-based format on request.

Capacity Management

On request, we will review the usage patterns across your equipment every 6 months from the date the equipment is commissioned for the purpose of monitoring the capacity of the equipment, for an additional charge. The charges for capacity management will be advised on application. Once we have completed the review, we will discuss the outcomes of that review with you (including any recommendations for equipment upgrades, configuration changes and additional capacity requirements in relation to conference ports and licensing), and agree any subsequent changes to your Managed Video Conferencing Service that may be required.

Change Management

You may request changes to your Managed Video Conferencing Service (being installations, moves, adds or changes or cancellations) by submitting a request in writing to us ("Service Change Request"), which we may accept or reject. If we reject a Service Change Request we will provide you with our reasons as to why it was rejected.

If we accept a Service Change Request, it will be performed in accordance with the terms set out in the Service Change Request. We will maintain a register of all Service Change Requests and make it available for review on request.

Reporting

On request, we will provide you with the following reports for an additional charge which will be advised on application:

- A monthly service report setting all faults and requests logged by the helpdesk;
- A daily system overview report which provides an overview of the status of your Video Conferencing system, setting out those endpoints and infrastructure that have triggered an alert;
- A monthly usage report; and
- A monthly telephone and queue report.

On request, we may provide customised reports for an additional charge which will be advised on application.

SUPPORT, MAINTENANCE, WARRANTY MANAGEMENT AND SERVICE LEVEL TARGETS

You may select one of the following service levels:

- Essentials Managed service – Business Hours;
- Essentials Managed service – Business Plus;
- Enhanced Managed service – Business Hours;
- Enhanced Managed service – Business Plus;
- Basic Managed service – Business Hours; or
- Basic Managed service – Business Plus.

The charges for your selected service level are set out in your application form or separate agreement with us. If you do not select a service level, your Managed Video Conferencing Service will be automatically provisioned, and charged for, with Essentials Managed service – Business Hours. You may select a service level independently for each video endpoint or video network device. However, if a video network device is deployed, you must select a service level for the video network device which is equal to or higher than that for any video endpoints.

These service levels are only available to you while your equipment is supported by us. We will notify you if your equipment becomes unsupported. In addition, the service levels are only available where the configuration and setup of your equipment is approved by us. If you modify our approved configuration or setup, these service levels will not apply to you. For unsupported or modified equipment, you acknowledge that helpdesk or on-site technical support staff may not have the relevant training, expertise or qualifications to assist with your queries. The table below sets out support available with each service level.

Support	Description	Essentials Managed service	Enhanced Managed service	Basic Managed service
Helpdesk	The helpdesk may be contacted via phone (an 1800 toll free number), video or email for all faults and requests in relation to your Video Conferencing service.	v	v	v
Technical Support (Phone, Video or Email)	The helpdesk provides technical assistance via phone, video or email.	v	v	v
Technical Support (Remote diagnostics)	Following a call to the helpdesk, technicians perform remote diagnostics by connecting to the suspect equipment, allowing deeper, more accurate investigation, remote rebooting and software fixes.	X	v	X
Onsite Fix / Replace	Technicians visit on–site to fix or replace equipment.	v	v	X
Equipment warranty management	We manage the warranty benefits for the equipment we provide to you.	v	v	v

Helpdesk and Technical Support

You can ask for our assistance with your Managed Video Conferencing Service via our helpdesk. The helpdesk will be available between 7am - 7pm (AEST or AEDST when applicable) on business days.

If you have a Business Plus service level, you may call the helpdesk 24 hours a day, 365 days a year. Between 7pm – 7am (AEST or AEDST when applicable) calls are directed to our afterhours call centre, which will pass your contact details to an on-call helpdesk representative who will return your call.

If you acquire equipment from us that allows for the use of peripheral devices and the peripheral devices are not purchased from us, you must contact the supplier for support of the peripheral devices. Our support is limited to equipment which is purchased, leased or rented from us, or which is otherwise approved by us.

Fault and Request Management

The helpdesk will make an initial assessment of the fault or request, and attempt to resolve the fault or complete the request via:

- Telephone, email or video conferencing; or
- Remote diagnostics (where available).

If the helpdesk is not able to resolve the fault or complete the request, we will provide you with on-site technical support during business hours. From time to time, we may engage subcontractors to perform on-site technical support on our behalf. You agree to allow our subcontractors to perform on-site technical support.

The helpdesk will:

- Log and provide a job reference number for each fault and request that you will need to reference when making any enquiries in relation to the fault or request;
- Track each fault and request through to resolution or completion;
- Provide regular status reports; and
- close faults and requests after confirming with you that the fault has been resolved or the request has been completed.

When contacting the helpdesk you must provide the following information:

- Contact details of your authorised contact person (including any site contact details and site address where on-site attendance is required); and
- Details of the fault (such as time of occurrence, symptoms and degree of impact) or request.

Web Portal

We may provide you with access to an online web portal which will provide you with access to reports, application forms, and other support services and tools in relation to your Managed Video Conferencing Service.

Alert Monitoring

On request, we will provide an Alert Monitoring service provided you have the necessary equipment to support the service. An additional charge will apply for the Alert Monitoring service which will be advised on application.

The Alert Monitoring service consists of:

- A status check for: basic IP connectivity of equipment; and system settings as agreed;
- An update of outcomes, including actions taken to resolve any issues identified in the status check;
- Basic network monitoring for packet loss, latency and jitter reporting (where available); and

- Backup of equipment database system configuration each month (however, if an external SQL database is deployed for the equipment, then the backup will need to be managed from SQL as part of an agreed process).

If an alert indicates a fault with the equipment, we will refer the fault to the helpdesk for resolution.

Exclusions

There are certain situations where we may charge you for on-site technical support. These include where:

- You have either Essentials Managed service – Business Hours or Enhanced Managed service – Business Hours and we perform on-site technical support outside of business hours on your request;
- You require on-site technical support as a result of an issue with your access service;
- You require on-site technical support as a result of a failure in your equipment and the failure is due to:

Your failure to follow our (or the equipment manufacturer's) installation, operation, maintenance or other reasonable instructions; Any unauthorised modification or alteration to the equipment; Any change or alteration to our equipment configurations, firmware or software; Abuse, misuse, negligent acts or omissions by you or any person under your control; or An event outside our reasonable control; You require on-site technical support on equipment that we do not support; or You require on-site technical support from us and you have a Basic Managed service (which does not include on-site fix and replacement of equipment).

Equipment Warranty Management

We manage the warranty benefits for the equipment we provide to you. We do not provide warranty management for equipment that you have acquired separately to your Managed Video Conferencing Service, even where the equipment has been approved by us for use with your Managed Video Conferencing Service.

Warranty management involves us liaising with the supplier of equipment on your behalf to report faults and seeking to obtain applicable warranty benefits such as repair or replacement of the faulty equipment. For example, if a part of your equipment is faulty and the warranty for your equipment specifies that you are entitled to a replacement part, we will liaise with the supplier to obtain this on your behalf. We will only provide warranty management during the warranty period for the equipment. The warranty period may vary depending on the equipment.

The procedure for warranty management is as follows:

- You must notify the helpdesk of the faulty equipment;

- We will notify the supplier of the equipment of the fault and find out from the supplier how we should proceed;
- We may direct you to send the faulty equipment to us or the supplier at your cost (if this is required by the supplier of the equipment);
- If you are entitled to a replacement part or the equipment is to be repaired, then the supplier will arrange for this through us and then send the equipment back to you; and
- A Telstra representative will install and commission the fixed or replaced equipment.

You acknowledge that warranty management is limited to us liaising with the supplier of the equipment on your behalf. We are not responsible for ensuring that you receive replacement parts or repair services and we are not responsible for the acts or omissions of the supplier of the equipment. You are responsible for all costs and repairs in your equipment for faults that are not covered by the applicable warranty. If you receive a replacement part, we (or the supplier of the equipment) may keep the part that is being replaced. If required, we (or the supplier of the equipment) may temporarily loan equipment to you. On request, you must promptly return the loan equipment to us at your cost and risk. Title to the loan equipment does not pass to you at any time. Risk in the loan equipment transfers to you on delivery.

Service Level Targets

We aim (but do not guarantee) to respond to requests regarding your Managed Video Conferencing Service within the time periods set out in the table below.

	Essentials Managed service - Business Hours / Enhanced Managed service – Business Hours	**Essentials Managed service – Business Plus / Enhanced Managed service – Business Plus**
Helpdesk Hours	Mon – Fri, 7am – 7pm (AEST or AEDST when applicable)	24x7
Response Time	120 Minutes	60 Minutes
Restore Times: Work undertaken during business hours only		
Urban Restore	End of next business day	12 Hours
Rural Restore	Urban + 1 business day	Urban + 1 business day
Remote Restore	Urban + 2 business days	Urban + 2 business days

Notes:

- All Restores times are subject to the underlying network availability and the subsequent restore time of those network services. The nominated restore times do not cover equipment failure requiring replacement equipment.

- Field time to restore for Remote sites can be improved if you hold spares at an appropriate location and acquire the appropriate training. Please discuss this option with your Telstra Account Representative.

Service Level Exclusions

We will not be liable for any failure to meet a service level where the failure is due to:

- An event outside our reasonable control;
- Any planned outages;
- Any acts or omissions of you (or your personnel), including any failure or delays to provide assistance or access to the sites; or
- Your network, systems, software or equipment that does not form part of your Managed Video Conferencing Service.

Planned Outages

From time to time we may need to implement a planned outage, which may involve us interrupting your Managed Video Conferencing Service to perform work such as critical network maintenance, system upgrades, modifications to hardware or software or testing.

We will use reasonable endeavours to:

- Provide you with at least 2 business days notice (via email or otherwise) prior to the planned outage;
- Ensure that planned outages occur on business days between 7:30pm - 6am (AEST or AEDST when applicable); and
- Ensure that any planned outage does not exceed 10 hours per quarter in total.

SPECIAL MEANINGS

The following words have the following special meanings: Business day means any day, other than a Saturday, Sunday or recognised public holiday in the state in which your sites are located. Business hours means 8am -6pm (AEST or AEDST when applicable) on business days. For the avoidance of doubt, if you have Essentials Business Hours or Enhanced Business Hours, access to the Helpdesk is between 7am and 7pm (AEST or AEDST when applicable), as set out in the Service level targets table above.

Equipment means video conferencing equipment. New Release means software which has been produced primarily to provide an extension, alteration, improvement or additional functionality to the Software, such as new versions.

Remote means a township or community with a population of less than 200 people. Response time means the time from which the incident is logged with us to the time when a resource is allocated to attend to the incident. Restore time means the time from which the incident is logged with us to the time when the incident is closed. Rural means a township or community with a population of 200 people or more but less than 10,000 people. Site means any location where the equipment is installed, or to which your Video Conferencing service is supplied. Software means any software we provide you as part of your Video Conferencing service, including any software embedded in the equipment. Updates means updates, bug fixes and patches to the Software, but does not include New Releases or versions of the Software or features or functionality which require additional licences. Urban means a township or community with a population of 10,000 people or more.

8

Video Equipment

VIDEO PRODUCTION EQUIPMENT CHOICES

Whether you're the designated family videomaker or an experienced video professional, you'll come to a point where you have projects that require an upgrade of your current acquisition system. (By acquisition system, we mean the gear that you take with you on the shoot, as opposed to the VCRs and edit controllers you may have in your editing system. When it's time for an upgrade, many beginners rush out to the nearest mall or video discount house, flash their credit cards and buy every new gadget they can think of. Without thinking ahead and planning their purchases, they end up with a closet full of expensive but useless stuff that won't work together and doesn't meet their videomaking needs. Professionals, however, don't just go out and buy a bunch of stuff. They purchase video acquisition systems. They make sure they have all the video, camera support, audio, lighting and accessories required to meet the needs of their video productions.

In this article, we'll define four major levels of video systems, from the casual shooter to the corporate/production video professional. Then we'll take a look at the different types of equipment that make up a workable video system for videomakers at each level. We'll toss out some specific products as representativ samples of what you'll find at each price point. With this information, you'll be able to make an informed choice about the videomaking equipment you need, rather than relying on a salesperson (or your own appetite for shopping) to choose your equipment for you. You'll end up with more than just a bunch of videomaking equipment; you'll have what the pros have--a video acquisition system.

LEVELS OF VIDEO PRODUCTION

Before you can even think about putting together a video system that best suits the level of production you want to achieve, you need to know what those levels are. You'll also need to know where your video projects fit into the overall videomaking picture. The four basic levels of sophistication in video production are the casual shooter, the serious amateur, the small-budget pro and the corporate/production video pro. At the first level is the casual video shooter. This level of sophistication includes those happy owners of the family camcorder as well as those who make a serious effort to document their vacations and family gatherings with steady video. If you're one of these, the gear you're likely to find in your closet or car is a fairly good but basic camcorder and a tripod.

The next level of sophistication, the serious amateur, consists of videomakers who spend their weekends doing simple video projects for their friends and family, as well as creating the occasional movie for their own artistic satisfaction. If this is your level, your current video kit probably includes a camcorder with a few special effects, a sturdy tripod, a camera-mounted or external mike, at least one light and a few other accessories thrown in for good measure. The small-budget professional probably owns a good 3-chip camcorder, a very solid tripod or other form of camera support, a wireless mike system and audio mixer, location lighting, gaffer supplies and a closet full of video accessories that have accumulated over the years.

The final level of sophistication for video production is the corporate/production video professional. This level includes those of you who produce corporate/industrial videos in a fairly well-equipped production facility. At this level, you should have at your fingertips a very good dockable small format camera, a color monitor, a high-quality tripod with fluid head and some other form of support such as a dolly, broadcast-quality mikes and mixers, full location lighting with a complete gaffer's kit and a utility dolly with a complete grip kit. Now that we've gone over the four levels of videomaking, let's discuss the first step in purchasing a new system: planning.

Starting Point

When putting together any level of video production system, you must base your decisions on two very important points: what are your video production needs, and how much can you afford to spend? When determining your production needs, ask yourself a few questions. Do you shoot most or all of your projects indoors, thus needing versatile lighting equipment? Do you mainly shoot outdoors, hiking to each location with only yourself or a friend acting as designated grip/pack mule? Do you make nature programs that require a lightweight yet rugged tripod with a very smooth fluid head and a camcorder with interchangeable lenses? Will interviews be a major part of your productions? If so, you may want to buy a soft light and other supplemental lighting equipment as well as a good wireless microphone system.

Your budget may help you decide whether you should buy the expensive wireless system or settle for a less expensive shotgun mike connected to the camcorder. With this in mind, read over the video production systems matrix. Keep in mind that every step up to a new level will more than double your investment in equipment and accessories. Take a hard look at the types of programs you would like to produce and the money you'll need to achieve the level of sophistication you desire; if your pocketbook can't accomodate the level of videomaking you d like to achieve, you may have some hard choices ahead.

Basic Principles

Once you have fixed firmly in your mind the type of video needs your system must address and the budget you must live with, you can then put together a video production system to meet those needs. The following basic principles will help you in making your decisions. The production system must include video, video support, audio, lighting and accessories. The first four categories are essential to the quality of your programs. The fifth category contains those things that make life easier but you probably don't need.

Many of us have seen the productions of those who buy expensive camcorders but forget the rest of the system. After watching only a few minutes, it feels like the final moments on a sinking ship; shaky, dark video that makes you seasick and audio that only the producer can understand. Good video productions start with solid, smooth video, quality audio and good lighting that matches the mood or style of the production. A video production system must provide all of these essentials. To get the most out of your system, buy individual components that are similar in quality, capability and cost. Shooting that important project with your new $1500 camcorder and a $50 tripod will be a very disappointing experience. It also makes little sense to use a $2300 tripod with a $600 basic camcorder. Choose your equipment so that each piece matches in quality and compatibility and complements the other parts of the system.

It is also very important to stay within your budget and buy only what you need. The price of a piece of equipment has very little to do with whether it will satisfy your production needs. Though the more expensive model may be a better piece of equipment, it may not provide the best bang for the buck. For example, a $300 wireless microphone system is usually better than a $50 system. It provides a better frequency response, higher quality audio and a greater operating range. A $2000 wireless microphone system may be even better than the $300 system--but not worth the extra $1700 to meet your specific needs.

Film directors and cinematographers are fond of saying "The quality of the picture begins with the quality of the piece of glass." When putting together your own system, you too should start with a "good piece of glass"--a quality camcorder, a sturdy, smooth camera support system, an audio system that gives good reproduction quality and lighting equipment that provides the luminance needed for the camera to produce a good picture. Once you have the solid basics and have paid for the quality your projects demand, you can look for inexpensive substitutes to handle the job of accessory equipment. The ability to do a lot with a little is always welcome. In lighting, you can often find substitute equipment and accessories. For example, instead of spending hundreds of dollars on a professional "butterfly" to light a large area with diffused light, build a frame of inexpensive PVC pipe and use lightweight tracing paper as the diffusing material. It's also important to plan your video system with the future in mind.

The more video production you do, the better your skills become. Better skills often lead to a desire for better, more sophisticated equipment. There will also be a greater demand for your skills. Unless you have an incredible nest egg or win the lottery, you won't be able to completely replace your current system. Build your new system one step at a time, improving one component of the system at each step.

Keep your needs in mind and buy to meet those needs. If you have a project that demands very smooth, steady camera movement, you may want to invest in a more expensive tripod even though it will probably be overkill for your current system. You should look at every purchase in terms of how the individual piece of equipment will complement the present and future equipment.

Assembling the System

Now that you have a firm grasp of your budget and you know what you want to spend your money on, you're ready to buy, right? Hold on. Before you rush out and damage too many credit cards or run up the company's equipment budget, look over the following suggestions and study the video production system matrix. These suggestions provide an overview of specific equipment models.

Most of the prices quoted are list and should give you a starting point for comparing the prices you find at your local video equipment outlet. As noted before, thematrixcontains basic information for each level of sophistication and provides ballpark figures for the actual cost (after shopping around) of buying a system.

Level One: The Casual Shooter

As we said before, the casual shooter only needs two basic pieces of equipment: a camcorder and tripod. The Panasonic PV-IQ505 VHS-C Palmcorder ($1000), RCA's CC422 full-size VHS camcorder($800) or Sony's 8mm CCD-TR65 ($1000) partnered with a Bilura Model 820 compact tripod ($75) would provide a quality basic system that is easy to use, provides a few nice features (all three have color viewfinders) and delivers good solid performance.

Level Two: The Serious Amateur

At this level, you'll want to be able to add some special effects to your productions and take your creativity to new heights. The JVC GR-AX75 VHS-C camcorder ($1200) or the 8mm Nikon VN-870 camcorder with a built-in color monitor ($1100) offer the convenience of compact size and a number of extra features. When buying a camcorder at this level, look for those features that will enhance your particular type of projects. You should be able to find features such as special effects, audio/video dubbing, manual override of all functions, wireless remote, character generator and color viewfinder. (For more information on camcorders and their features, refer to the December 1994 Videomaker camcorder buyer's guide.) You may want to include the Citizen M329 full-color video monitor that slides into the accessory shoe of your camcorder ($250). An external monitor like this one or the one found on the 8mm Nikon VN-870 camcorder offers a great deal of flexibility to your system. You no longer have to strain your eyes to compose your shot and the color monitor provides an instant check of white balance and lighting contrast.

No matter how fancy your camera is, if you don't provide solid, smooth support, your video will still look as if you're shooting it with a $500 blue-light special camcorder. The Bogen 3021 tripod ($146) with 3130 head ($85) offers rock-solid support with smooth pans, a quick-release plate and the ability to adjust the height from 10 1/4 to 71 inches. If you shoot on the go, you may want to opt for Sima Products Corporation's Mini VideoProp ($40), a support that rests on the user's chest, using the body as the tripod. You might also consider a monopod such as Bogen's BTH-3249 ($50). We sometimes think of audio as video's little brother, but it's an integral part of the production. Bad audio will destroy a video production even if the video is awesome. To supplement your camcorder's audio system, you may want to look at the Crown Sound Grabber ($99). This PZM-style area mike has a long cable and works with any consumer camcorder. Another direction you may go is the Nady VCM-100 Camera Mount Microphone ($65). To hear your improved audio during recording, pick up a set of headphones with a mini 1/8-inch connector ($30-$80).

If you provide enough light and reduce the contrast between the dark and light areas in a picture, VHS and 8mm tape can produce some amazing results. For supplemental lighting, consider the Sun-Pak CV-20s Video Light ($100 for the kit, including battery and charger). For accessories, the serious amateur should consider extra batteries for the camcorder and light ($35-$150 depending on the equipment). Finally, a gadget bag ($35-$100) will make getting your equipment around a lot easier.

Level Three

The Small-budget Professional. This level has become the focus of a great number of equipment manufacturers and has the widest range of pricing. Aimed directly at this level of production is the "prosumer" line of equipment that you can find in almost any video equipment and supply house. Your camcorder choice should be either an S-VHS, S-VHS-C or Hi8 model with over 400 lines of resolution. Knowing your budget and production needs is critical in buying a camcorder at this level. In the S-VHS or S-VHS-C format you may choose the Panasonic AG-456U ($2495) or the Panasonic AG-3 camcorder ($3570). In the Hi8 format, consider the Hitachi VM-H71A ($1900) or Sony's CCD-VX3 ($3800). If your productions depend on interchangeable lenses, you will want to take a look at Canon's L2 ($4000).

At this level, you should include alternatives to the standard tripod camera support. One such alternative is the Steadicam JR ($500). This small version of the professional Steadicam, while a little difficult to master, is well worth the effort. It provides smooth camera support on the move. The Bogen Deluxe Video Dolly ($164) is another addition to your camera support system that you may want to consider. As your video improves, so must your audio. For interviews, you'll need a good wireless lapel mike system like the Nady 351 ($299). The Shure M267 professional mixer ($535) can be battery powered to handle your mixing needs in the field. You will also want quality headphones such as Sony's MDR-7502 ($62).

Clear, well-lit video is the sign of professional. The Lowel Standard Softlight 2 ($825) provides a soft, controlled light that will enhance your interviews and any location shoot. To supplement the Softlight, consider an NRG two-light Photoflood kit ($340). Don't forget to add a lighting kit that includes Gaffer's tape ($12 per roll), 3x4 foot sheets of white foamcore for bouncing light, black poster board for flags, 3-pin to 2-pin AC adapters, extra extension cords, a power strip, clothespins and any of your other favorite lighting accessories. One item that the low-budget professional will find very useful is a video cart. Most professionals that work at this level specialize in being a one-man band. The Bretford BPDUO-EW ($205) has a big wheel assembly for smooth transportation of your camera, tripod, lights and everything else. It also has a built-in 4-plug power strip and 20 foot cord. If you need to travel, it comes apart quickly and fits neatly in a van or car trunk.

Level Four

The Corporate/Production Video Professional. You will still find S-VHS, S-VHS-C and Hi8 at this level of production. However, the level of sophistication in the cameras is greatly enhanced. The camcorders at this level include Panasonic's S-VHS format AG-DP800 ($9900). This level also includes cameras with separate "dockable" recorders such as the JVC KY-27U camera with dockable BR-S422U S-VHS recorder ($15,000) and the Sony DXC-325 camera with the EVW-9000 Hi8 dockable recorder ($9000). If your budget can afford a dockable unit, it will be more adaptable in the future should new and higher quality video formats become a financial possibility. The larger cameras found at this level require more substantial video support. The Vinten Vision System 5 tripod and fluid head ($3295) is a sturdy lightweight example of such a system. The Vinten Vision 3319-3B castoring dolly ($635) will add even more flexibility to your remote shooting.

For audio at this level, you need to purchase broadcast-quality mikes. The Audio-Technica AT815a shotgun ($300) is a good all-around location mike. For voice and handheld work, the Shure SM58 ($215) is a solid performer. A high-quality wireless system and production mixer are also part of the formula for successful audio at this level.

You may want to supplement your production capabilities in audio with a portable audio recorder. The Marantz PMD 430 portable stereo cassette recorder ($599) is excellent as a recorder for ambient sound and sound effects. The corporate/production professional usually has an array of lighting kits for various location setups.

The Lowel Ambi kit ($1695) is a very versatile kit, usable in a number of situations. The lighting budget should also include items such as flags, gels, reflectors and C-stands for holding it all. Finally, the closets and desks at this level are usually overflowing with all kinds of accessories that simplify life and add to the video professional's abilities. One accessory that will make life easier is a van. If you're on good terms with an auto dealership, you may try to spring for a production van. But, until they hand you the keys, load up your video cart and enjoy your videomaking experience.

Final Thoughts

Having lots of cash, a van full of equipment and a dream budget does not guarantee super productions. You must put a great deal of time and effort into the preproduction (planning, budgeting, scriptwriting) portion of the videomaking process. If you plan your productions well, keeping in mind your equipment's limitations, you will be successful.

BASIC GUIDE TO SHOOTING VIDEO

This paper looks at practical concerns and techniques used in the shooting of video. The intention is to give the beginner tips and tools as well as to highlight some common pitfalls the beginner may encounter. This document is intended to give the reader a brief introduction into techniques for making videos.

As many of the issues relating to exposure, composition, etc. are identical to those encountered with still images, this document will focus on issues which are particular to moving images. Similarly, only the briefest mention will be made of audio concerns, these being covered in documents written for audio recording and also in Audio/Video Production: Recording Lectures, Seminars and Events.

EQUIPMENT

Cameras

It is important to consider the final form the video will take. If you are shooting something that will appear on a website it is a waste of time and money to use a DigiBeta camera. Conversely, shooting footage on a PD-150 will not produce footage which is generally considered suitable for projection in a cinema (although this is, in fact, exactly what David Lynch did for his film Inland Empire).

This also applies to video formats. Many cameras offer a choice of format, such as DV,DVCAM and HDV. Again, it is useful to consider the end purpose of the video footage. If the only end product is going to be a low-resolution web video clip, then it is a waste of time, money and resources to shoot it in a high-definition format such as HDV. On the other hand, if the footage is one where a very high-resolution image is required or if it is possible that the footage may be used again in the future for an as-yet unspecified purpose, it may be worthwhile to shoot it in as high-resolution format as possible.

Tripods

After the camcorder itself, your single most important piece of equipment is a good tripod. What makes a good tripod? In a word, weight. A light, small, portable tripod is next to worthless for video work because it is largely incapable of doing the one thing it is intended for: holding the camera steady. This doesn't mean that you have to lumber yourself with an enormous dead weight in order to shoot reasonable videos. A good, solid tripod for a typical consumer or prosumer camcorder will weigh only a couple of kilograms.

The head of the tripod is equally important and it is here that video tripods and stills camera tripods part company. Unlike a stills camera, a video camera records movement: it will be tilted (changing its up-and-down orientation) and panned (changing its left-and-right orientation) while it is shooting and it is essential that these movements be smooth and deliberate. While a typical stills camera tripod head will allow such movement, its use will not result in smooth movement. For that one needs a video head. There are innumerable video tripod heads available. It is important to get a head which is rated for the weight of camcorder you are using. The best heads for video work are fluid or semi-fluid heads. These (particularly the true fluid heads) ensure that any camera motion will be smooth.

Microphones (on-camera mics vs. Separate mics) and Headphones

We will not go into detail about microphones here, as this subject is already covered in theMicrophone Guide and Microphone Technique articles. There are two points we will make, however. First, if at all possible the use of built-in mics should be avoided. Not only are such microphones generally of a relatively low quality, but there is also the problem of camera noise. Any noise made by the tape or disc transport mechanism, the zoom lens and the aperture (not to mention the camera operator) can and most likely will be picked up by the camera mic. Second, it is vital that closed-back headphones be worn whenever audio is being recorded that will be used in the final video, as this is the only way to ensure that useful audio is being recorded.

PREPARATION

Purpose of the Video

Before embarking upon a shoot it is vital to consider what the purpose of the shoot is. This is important for several reasons. First of all, it may be that the project doesn't actually need video. Assuming it does, one should determine how video will help the success of the project. Should it be a demonstration or a lecture or perhaps a workshop session? Should it be published on the Internet or used in a PowerPoint or distributed as a DVD? Even if no clear answers to these questions can be found yet, the exercise of considering such issues can only help with decisions that must subsequently be made.

Tape and Battery Eequirements

Care should be taken to ensure that sufficient tapes (or discs, or whatever medium is being used) and batteries are taken on a shoot. In the case of tapes, it is particularly important to check the format that the video is being shot in. For example, the same tape which lasts an hour when shooting in DV will only last 40 minutes when shooting in DVCAM. It is good practice to have sufficient tapes to cover all of the anticipated filming plus at least one extra tape per camera used. Although highly unlikely, it is possible that a tape could jam or be otherwise damaged.

In the case of batteries, always ensure that they are fully charged the night before a shoot. While it is possible (and a very good idea) to get larger-capacity batteries for camera that will cover an entire day's shoot, it is always a good idea to bring along at least one spare battery and the charger - and an extension lead. Remember, too, that you will often be in a situation where the camera can be run off the mains, allowing you to save the battery for when it is really needed.

Labelling

Tapes (or whatever medium you are using) should always be labelled at the shoot, preferably before they are used. The label should include the name of the shoot, the date, information about how sound is being recorded (camera mic, boom, line in, etc.) and the tape number (it is worthwhile to number the tapes "tape 1 of," "tape 2 of" and so on, filling in the total number of tapes used at the end of the shoot).

Optionally, one can also add the name of the camera operator and any other information that may be useful in post-production - if there's room on the label, of course...

Psych lecture
23 September 2008
Ch 1: line in from PA
Ch 2: cam mic
Tape 1 of 3
Student interviews
3 October 07
Ch 1 & 2: hand-held mic
Camera: Jane Doe
Tape 2 of 2

The Get in and the Get out (considerations, Health and Safety)

Remember when setting up for a shoot that you may want to go back to this location some time in the future, so leave a good impression. Get permission for the location you are using, any furniture or other equipment you need and access to electricity. When you are finished return the location to the state it was in before you arrived (or better).

Fig. Typical plug-in RCD

Health and safety concerns should be paramount. All electrical equipment including leads and extension leads should be PAT tested. It is a good idea, particularly when using lighting, to bring along a plug-in RCD (residual current device) and run the extension lead through it; in the case of faulty wiring on a light it may save someone a nasty shock or worse. All cables should be taped down with gaffer tape to eliminate trip hazards. Cameras and other equipment must not block fire exits. And don't leave anything unattended that you don't want stolen.

Setting up the Camera

Some video cameras allow the user to select the audio sampling rate, generally either 32000 Hz or 48000 Hz. The former will result in a small saving of disk space if the video is transferred to a computer for editing. On some cameras it allows the tape to hold 4 channels of audio instead of 2, but it will also record the audio at a slightly lower quality. However the difference in quality is negligible. On the other hand, the extra disk space consumed by recording at 48000 Hz is also negligible, and while the tape can hold 4 channels, the camera will only record 2 at a time - and 2 channels is enough for most purposes. Probably the best approach is to pick one sampling rate and stick with it, particularly as some older editing programs don't allow you to mix sampling rates.

Many video cameras also allow the user to set up zebra patterning. When this is used, areas of the image in the viewfinder or viewscreen which are exposed above a certain point are filled with a striped pattern. If the camera is set to autoexposure, there is no point in having zebra patterning. However, this can simplify the process of getting a correct exposure when it is being manually set. Like any such guide, it is necessary to experiment and practise with zebra patterning in order to benefit from its employment.

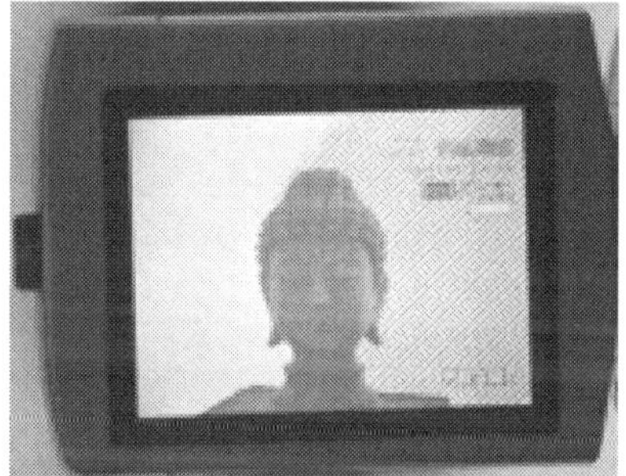

Fig. The right of the screen has a zebra pattern, indicating a very bright portion of the image

Image stabilisation is a very useful feature included in many camcorders. It can significantly increase the sharpness of handheld shots but is not necessary if the camera is mounted on a tripod. In addition, all image stabilisation systems are incapable of distinguishing between a jiggled camera and the beginning of a panning movement. As a result, pans may appear jerky as they commence, although newer camcorders have improved software to eliminate much of this jerkiness. On balance, if you are shooting handheld and do not intend to do a lot of panning, image stabilisation is probably a good feature to use - but it is always best to experiment with your equipment beforehand.

PRACTICE

Holding/Steadying the Camera with a Tripod

An essential part of shooting video is ensuring that the camera is steady and camera movement is smooth. This is easily accomplished with a tripod. Most modern video tripods include a spirit level in the tripod head. This is a very useful piece of equipment; in fact, if your tripod does not have a spirit level it may be useful to get a small one for your video kit. This is because the eye is easily fooled about what is horizontal, particularly on an uneven surface. In addition, a panning shot that looks level at the start may end up with a slant at the end if the tripod isn't level front and back as well as left and right.

The smoothness of camera motion on a tripod is affected by how you hold the camera and the panning arm. Most importantly, the further from the camera you hold the panning arm, the smoother the movement will be, as any jerkiness will be reduced. It is also often a good idea when using the viewfinder to place some of your weight on the camera by resting your hand on top of it. Your weight will serve to dampen any unwanted motion. In addition, on many cameras you can start and stop the camera and control the zoom with your left hand while panning with your right.

Holding/Steadying the Camera without a Tripod

A tripod is not always available or practical for shooting. There may not be enough room to set it up or the shot may need to be lower or higher than the tripod can reach. Perhaps the camera has to be in several different places and the tripod can't be carried around with it. While tripods are always the best way to get a steady image, they are not the only way. With a little practice it is possible to get quite good results without a tripod, either as a hand-held camera or using alternatives to a tripod.

For static shots, the best practice is always to place the camera on something. Almost all camcorders are designed to stand upright on flat surfaces, so any flat, stable surface can be considered as a place to put the camera: tabletops, chairs, low walls, ladders, even the floor. If the surface is not even, place a wedge, a small stone or a tape case under the camera to level it - or shoot the image at an angle. Some camerapersons include a beanbag in their kit for the express purpose of levelling the camera. A coat or jumper can also be very useful for this.

When a level surface of the appropriate height is not available or when you need to pan or tilt the camera it is necessary to hand-hold it. For maximum stability you should follow a few key points. First of all, your legs should be comfortably spread, about a shoulder width apart. You should try and adopt a relaxed posture, not tense, but not floppy either.

Use both arms to hold the camera: place your left hand, palm up and fingers facing forward, under the camera to support its weight (if the camera is too small for this, wrap your left hand around your right hand just above the wrist). The right hand is placed through the camera strap (if there is one) to add further stability and to operate the controls. Most importantly, keep your elbows tucked in against your ribs. This is a position which initially may feel unnatural but which adds tremendous stability to your shots. In particular, resting your elbows against your ribs means that your upper arm muscles are not supporting the camera, a situation where they will quickly tire. This is important not only for your comfort, but because as muscles become fatigued they start to introduce small movements.

The most stable support for anything is a tripod, so, if possible, make yourself into one. Two of the tripod's legs will be your own. The third can be provided by a wall, a pillar or a tree which you can lean your back against. You will be able to hold the camera more steadily and once again your muscles will not tire as quickly. The most important thing to remember about panning is not to do it too quickly. A pan which feels quite leisurely when you shoot it can appear quite rushed when you view it afterwards. There is a technique called a whip pan, where the pan is intentionally executed as quickly as possible in order to provide a place to cut to another scene or simply to disorient the viewer, but it is rarely used.

One should also consider the focal length of the lens when doing a pan. A camera movement which, when the zoom is set to wide angle, results in a slow pan will result in an extremely quick pan when the lens is set to telephoto.

The next most important thing to remember about panning is your body position. The normal tendency for the novice cameraperson is to stand facing the position where the pan begins and then to twist the body as the camera pans until you reach the end of the pan. Panning this way means that, as the pan ends, the body is twisted and as a result very likely to introduce shake into the image.The correct way to pan is to position the body so it faces in the direction the camera will point at the end of the pan, and then twist it back to the start position of the pan. This way the panning movement is an untwisting of the body, which is then maximally relaxed when the pan concludes.

If, however, the cameraperson's start position is photo 'b' and his end position is photo 'a' then he is using bad panning technique: the body twists and tightens as the pan progresses, making it more difficult not to shake the camera or pan it jerkily. A useful and inexpensive piece of kit which can improve your results significantly is a string tripod. This is simply a bolt which fits into the threaded hole in the base of the camera with a loop of string which is long enough to reach from the camera in its shooting position down to the ground. By placing your foot in the loop you can provide something for your arms to work against, making the camera more stable. In particular, with a little practice you can use this to get very stable and pure panning movements.

Viewscreen vs. Viewfinder

Several brief points should be made about the relative merits of viewfinders and viewscreens. Viewscreens enable the camera operator to move the camera away from the face: cameras can be held over the head or down at knee level and the operator can still observe the image that is being recorded.

Viewfinders, however, still have their advantages. If manual focus is being used the viewfinder provides a more reliable indicator of how sharp the image is. Furthermore there is the sometimes critical point that the viewfinder consumes much less power than a viewscreen. In cases where battery life is a concern the viewfinder may mean the difference between success and failure.

Autofocus

All modern camcorders have autofocus. While this facility greatly simplifies the shooting of video, there are four types of behaviour associated with it which can cause problems. The first is due to the fact that the autofocus mechanism is designed to focus on whatever is in the centre of the frame. It is quite common when filming lectures or interviews for the subject of the shot not to be centred in the frame. In this situation the autofocus may focus on the wall behind the subject rather than the subject, resulting in an out-of-focus shot. Still cameras provide a means of dealing with this by allowing the user to let the camera focus on the subject and then lock the focus at that point, after which they can compose the shot.

Camcorders don't generally have this facility. Instead they have a means of easily turning the autofocus on and off. Using this is much like the lock focus on a still camera: focus on the subject, then turn off autofocus and compose the shot. The key difference is that the autofocus must be turned on again afterwards. The second type of behaviour which can cause problems is related to the first. Because the autofocus adjusts itself based on what is in the centre of frame, during a shot which involves camera movement the focus will change. Sometimes this change of focus is welcome, such as when the camera pans from one person to another or from a speaker to a screen. When it is not wanted, autofocus can again be turned off, maintaining the original focus.

The third behaviour which can cause problems occurs when the camera is static on a subject, but the focus keeps changing slightly. Occasionally the autofocus gets into a situation like this where it continually re-adjusts the focus. Again, switching off the autofocus is the only way to deal with this. Finally the situation occasionally arises where the autofocus fails to respond. The subject is noticeable out of focus and it remains that way. In a case like this the simplest way to re-establish focus is to pan the camera away to an object a different distance from the lens, allow the autofocus to focus on it and then pan back to the subject. If this fails, turning the autofocus off and on again should correct the problem.

Of course, none of these problems occur if the camera is set on manual focus. This is what professional camerapersons do, and while it takes more work and precision on the part of the operator, it is a skill which can be learned quite quickly. The most important thing to remember if you are focusing manually is to zoom into your subject before you focus and zoom out to the frame you want afterwards: it is much easier to focus when zoomed in and much more precise, due to the greatly reduced depth of field of the lens. There is one caveat to this procedure, however. While almost all modern zoom lenses will keep in focus throughout the full range of the zoom, some will not, Be sure to test your camcorder to ensure that a sharp focus when zoomed in will be maintained when zoomed out.

Autoexposure

As with autofocus, autoexposure can result in behaviour that may be problematic, although this is less likely. When changing the framing of the camera the exposure will change, but changes in exposure, unless extreme, are much less noticeable. A more common problem is one where the exposure needed to correctly see the subject differs from the exposure that autoexposure selects for the image. The most common example of this is when a person stands in front of a window or a screen and as a result appears as a silhouette.

There are two ways of dealing with this. Many camcorders have a switch (sometimes labelled "backlight") which is intended to deal specifically with this issue. Depressing the switch opens the aperture a stop or two above what the autoexposure has set. This will improve exposure of the subject while overexposing the background. However, this may not be the exact increase in aperture needed for a correct exposure. Best results can be attained by turning off autoexposure and adjusting the aperture manually. On some camcorders the shutter speed selected by the autoexposure results in a strobing effect when recording camera movement or a rapidly moving subject which gives all such movement a jerky look. While this is arguably a bad way to record images, its frequent appearance in videos has resulted in it becoming an accepted style of shooting video. It can be eliminated by selecting a fixed shutter speed on the camcorder.

Composition

The quality of the images you record on a camcorder can be significantly improved by remembering a few simple points. First of all is to be aware of whether or not the camera is level from left to right. This is not to say that it should always be level; rather, one should ensure that the level shots are truly level and the non-level shots are quite clearly non-level. A shot which is only a bit out of level generally looks sloppy. Remember, if your tripod head doesn't have a spirit level built into it, a small one can be purchased at most hardware stores. This idea of making things look deliberate also extends to camera movement. If the camera is supposed to be still, keep it still; if it is supposed to move, make the movement clear, smooth and deliberate. There was a fashion in several television series a few years ago to have the camera continually drifting around, emphasising that it was being shot with a hand-held camera. This effect can be annoying and is most certainly distracting.

As always there is a counter-example to this practice. It is common when shooting someone speaking to make sure that the camera is not locked in position on the tripod. This is so that the cameraperson can correct the framing when the person speaking moves, as speakers almost always do. Good composition requires that you think about the camera's position relative to what's being shot - and what's happening around you. If you are videoing an interview with someone sitting in a chair you don't want the camera shooting from a standing level, looking down on the interviewee. If you are recording a lecture you may want to position to get a student's-eye view - but at the same time you don't want to block the view of any students attending the lecture.

Consider, too what else is in the frame. The presence of a legible poster located behind the person you are videoing may tempt the viewer to read it rather than listen to what is being said. Similarly, if some sort of action is being carried out in the background which isn't directly relevant to what is being said, it could prove distracting. A classic error is to shoot a person such that it looks like a tree or a building is growing out of the top of their head.

Even the choice of focal length will affect the image dramatically. A person on a street shot with a long telephoto can look trapped or lost in crowds of other people, even when the street is fairly empty. Conversely, nothing can destroy the impressiveness of a mountain range or a tall building like a very wide angle lens. However, the most important guideline for the beginning cameraperson can be summed up in two words: more close-ups. Beginning camerapersons tend to include too much in their compositions, just as beginning scriptwriters include too much in their scenes. By keeping the information in the frame to a minimum we minimise distractions and confusion for the viewer.

The cameraperson should consider where the video will be viewed. Simply put, the smaller the screen, the bigger the objects should appear in frame. It is common practice in the film and video industry to shoot for television differently than for cinema. In particular, television shooting includes far more close-ups and detailed shots of objects. This is even more important in this day of streaming video and mobile devices where the video may be viewed on very small screens. Try and keep the compositions interesting. A speaker should almost never be shot speaking in the direction of the camera: it is far better to have them facing, however slightly, to one side or the other. Placing a speaker in the centre of frame is both rather boring and unbalanced looking. It is much better to place them off-centre so that there is empty space on the side of the frame they are addressing.

Crossing the Line

There is a fundamental concept in the shooting of film which is referred to as "crossing the line." This concerns the restrictions we must place on how we shoot a scene so that we do not give the viewer contradictory information about the spatial relationships in the scene. If we imagine the scene we are shooting as taking place on a stage, "the line" can be thought of as the front of the stage. As long as we stay on the audience side of the line there will be no confusion about the positions of everything. As soon as we cross the line, however, the viewer will receive information contrary to what has been shown before. In the case of a scene with two people speaking, the line can be thought of as passing through or just in front of them. Sometimes we have a choice in where we place the line. This is fine, so long as we choose one position and stick to it.

There are always situations where any rule can be broken; so it goes with this rule. Finally, a counter-example and a bit of trivia. In the 1939 film Stagecoach, the scene where the coach is attacked is famous for being a textbook example of how not to preserve spatial relationships. The line is crossed again and again and the stagecoach is continually changing direction as a result – and it doesn't matter, because the viewer is so caught up in the action that the discontinuities aren't really noticed.

Shot Selection

One of the most common errors novices make is to move the camera too much or too soon. People tend to get restless when shooting video and have a very different impression of the passing of time to the people who eventually view it. In a typical situation the camera operator takes a shot, holds it for what seems like an eternity and them moves to another shot, but when the footage is subsequently viewed it seems that that camera was hardly at rest before it moved on to something else. This is, if anything, even more critical when the footage is to be edited, because the beginning and end of the shot will almost certainly have to be discarded.

A good rule of thumb is to frame the image you want to record and count to 10 before moving on to another shot. Obviously this cannot be a hard and fast rule: if you are capturing a performance, for example, you may suddenly need to shift your attention. But in many cases, such as capturing images to be edited into a sequence later or taking shots of objects, the 10-second rule can be invaluable. Another area where people move too quickly is at the beginning and end of the shot. Having set the camera to record, it is good practice to wait a few seconds after the red record light is on before getting the person on camera to speak. Similarly the camera shouldn't be turned off the instant the person has stopped speaking.

One should also beware of overlapping speech. If your subject starts talking a fraction of a second after you have told him or her to, it may be difficult to remove your words from the shot without cutting off the subject's. If you are intending to edit your footage it is essential to shoot cutaways. These are shots which can be used to cover edits in the main footage. For example, you may be shooting a lecture. Before or after the lecture you take some shots of people in the audience listening or simply looking towards the front of the lecture hall. When you edit, if you want to cut out a bit of the lecturer's talk, you can switch the picture to the cutaway at the point where you cut the lecturer's speech and, if properly done, the viewer will not realise that an edit in the speech has occurred.

Videoing Supporting Materials

When videoing an event it may be necessary to shoot all the footage one needs as it is happening. For example, if shooting a lecture it may be necessary to move from the lecturer to a projected image and back throughout a lecture. However, if the footage is to be edited and time allows, it is far better to shoot supporting materials later and cut them into the finished video. Not only does this mean that there is less chance of missing important shots (in our example, it means that the video could record the lecturer throughout the lecture), but also that the supporting materials can then be shot under better conditions. For example, instead of shooting a large screen, one may shoot the supporting materials on a computer monitor or even shoot paper copies of the materials.

With a little care excellent results can be achieved with a standard video camera and no other equipment. Sheets of paper may be shot on a lectern or even stuck to a wall. Provided the light is adequate and the camera is locked off on a tripod the results will be more than adequate. If shooting an image from a data projector it is vital that the projector be focused: remember that an out-of focus image can't be re-focused in post-production. This is even more important in this situation where the focus of both the data projector and of the video camera can contribute to a blurred image.

Focusing is less of a problem when shooting a computer monitor but there is a caveat: while one may shoot a modern LCD monitor with no problems, the same is not true of older CRTmonitors or televisions. Because of the way such screens work, shooting will frequently result in a horizontal black band across the image which, in worst cases, may slowly descend down the screen.

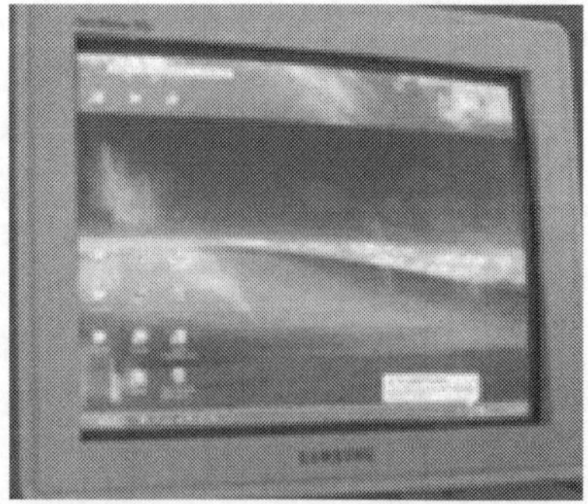

Fig. An example of the black band that appears when videoing a television or CRT monitor.

The correct way to eliminate this problem is to bring in an expensive piece of kit called a scan converter. The best way to eliminate this problem is never to shoot a CRT monitor or TV. However, if there is no alternative, there are two possible ways of dealing with this problem. The first method is simple and 100% effective, but is only possible with camcorders which permit the user to vary the shutter speed. On such devices one need only set the shutter speed to 1/50 of a second or less (1/60 in the case of an NTSC monitor of the sort found in North America).

The second method is less reliable but should produce good results with a little patience. Where the black band appears on the screen is random, depending on when in the monitor's refresh cycle the camera is switched to record. Each time the camera is turned on, the black band will appear in a different place. Sometimes the band will be off-screen. As a result, if one keeps switching the camera from record to stand-by and back to record, eventually the image in the viewfinder will show no black band and the recording can proceed. The recorded image should be closely watched because the black band may begin to appear at the top of the screen. In this case it will be necessary to switch the camera to stand-by and once again attempt to get the band off screen. This can be fiddly and frustrating, but in the absence of an alternative it will usually work.

Loose Ends

When you finish recording on a tape, if you do nothing else it is essential to write protect the tape. This is normally performed by sliding a small piece of plastic to uncover a hole in the tape cassette. If you make it a habit to do this as you remove the tape you will go a long way to ensuring that you never accidentally erase footage you have shot.

Fig. The write protect tab on a DV tape: The tab should be in the position in the top photo when it goes into the camera prior to the recording, but should immediately be moved to the position in the bottom picture when it is taken out.

For the purposes of editing the footage it is always a good idea to record what is known as "atmos" in the filmmaking world. This is simply the background noise in a room. It could just be the ambient noise or the sound of the air conditioning, but it could also be traffic sounds coming from outside or the sounds of students talking.

This sound is useful to ensure that there is uniform background noise in an edited piece of film. It is normal practice to record 30 seconds of atmos in any given location. If you need longer than this in the edit it is a simple matter to re-use the same 30 seconds: this is a long enough time that the listener won't realise that the sound is being repeated. Don't be afraid to shoot re-takes.

You may find that the footage you have taken simply is not good enough. It may be possible to re-shoot the material immediately, but there is no reason not to shoot it days or even months later provided there are no problems with continuity: a lecturer whose hair and clothing change from shot to shot will be, to say the least, distracting.

EQUIPMENT, INSTRUCTIONS AND IDEAS FOR MAKING VIDEOS

Ready to make great videos? Learn about the equipment, tools and accessories you'll need to start making videos. Pick up tips for putting that equipment to use, so you can make videos that look and sound great. Get ideas and instructions for different types of video to make.

VIDEO EQUIPMENT

You don't need expensive equipment to make videos, but having the right stuff helps. A good camcorder, editing computer and accessories will improve the quality and ease of all your video-making.

Choose the Right Camcorder

A camcorder can be a big investment, and it is one of the most important pieces of equipment when it comes to creating high-quality desktop video. With the abundance of video cameras available, it can be difficult to figure out how to buy a video camera that's right for you. Before you run out and buy the latest model or whatever is on sale this week, consider the following tips. They're designed to help you figure out how to buy a camcorder that's perfect for your needs.

Consider Your Needs

First, consider what you'll be using the camcorder for. If you have dreams of being a YouTube star or starting a vlog, you should buy a camcorder with a webcam function. If you're going to be bringing the video camera to family events and on vacation, you should look for something small, light and easy to use. If you're going to be shooting your big movie or doing professional work, you'll want to spend a little bit more to get higher end features and better quality.

Features

Once you have an idea of what you'll be doing with your camcorder, you should be able narrow down the features that you'll need.

Format

High definition or standard definition, that is the question. You can find many affordable high-definition camcorders now, but prices are still dropping as quality improves. If you want to be on the cutting edge with high-definition, remember that besides an HD camcorder you'll need HD editing software, an HD-DVD or Blu-ray burner, and an HD-DVD or Blu-ray player.

Recording Medium

Most standard definition consumer camcorders record either to mini-DV tape, DVD or hard disk. Each medium has its advantages and disadvantages.

Accessories

What comes included with your camcorder, and how much do additional accessories cost? Remember to consider the cost of batteries, microphones, cables, memory cards and editing software when factoring the price of a camcorder. If some or all of these things come included, it increases the value of your purchase.

Do Your Research

Once you've narrowed down your camcorder choices, read some reviews to find out more information about the models you're thinking about purchasing. These reviews might point out information you haven't considered, or reinforce your purchase decision.

Choose the Right Microphone

If you want to record high-quality audio, you shouldn't rely on the built in camcorder microphone. Instead, you should use an external mic for your video camera, which will pick up sounds more clearly and accurately.

But buying an external mic for your video camera can be a tricky proposition: you're faced with lots of choices, and it can sometimes be difficult to make a decision. These tips are designed to help you pick out an external camcorder microphone.

External Mic Connections

The camcorder microphone you buy will be dictated by the type of external mic connection built into your video camera. Consumer camcorders often have a stereo jack for attaching an external mic, while higher-end camcorders will have an XLR jack for connecting a mic. Before you buy an external microphone, be sure to check what kind of input your camcorder has, and pick out a microphone that will fit the jack. You can also visit your local electronics store and buy a camcorder microphone adapter, which will allow you to connect most any external mic to the input jack on your camcorder.

Types of Camcorder Microphones

There are three main types of camcorder microphones to choose from: shotgun, lapel (or lavaliere) and handheld (like newscasters or musicians use). Each type of external mic is suited to a different type of video production, and ideally you'll be able to purchase one of each kind.

Shotgun Microphones

Shotgun camcorder microphones can be mounted on your camcorder or attached to a boom pole. The microphone will pick up all of the sound coming from the general direction in which it is pointed. Shotgun camcorder microphones work well for video productions in which you want to record ambient sound or audio coming form multiple speakers.

Lapel Microphones

Lapel microphones are great for video interviews. You attach them to the subject's shirt, and they'll pick up the person's voice very clearly, as well as any sound that is close to the mic. Lapel microphones are also very useful when recording wedding videos.

Handheld Microphones

Handheld microphones are usually pretty heavy duty and durable. They work great for picking up nearby sound (so your subjects need to speak right into them).

However, they definitely lend a very "newsy" look to your video, so they're best used if you're going for that newscaster look, or if the speaker is not going to be seen on camera.

Wired and Wireless External Mics

Most types of camcorder microphones are available in wired and wireless versions. Wired camcorder microphones connect directly into your camera. Wireless microphones, on the other hand, come with a receiver and a transmitter. The transmitter is connected to the microphone, and the receiver is connected to your camcorder. Wireless camcorder microphones are very handy because you can record audio that is very far away from your camera.

However, they are also much more expensive than wired microphones, and you have to take into considerations things like range, signal interference and battery power.

Camcorder Microphone Quality

Once you've decided on the type of camcorder microphone that you're going to buy, you still have to choose a make and a model. There is no one external mic that's best for everyone, so you'll have to do some research to find one that fits your needs and budget. Read reviews, talk to video producers, and get your hands on as many camcorder microphones as possible so that you can hear the audio quality for yourself. Invest in a quality external mic now, and you'll be able to use it for years down the road. Whether you're shooting in HD or for the internet, a good camcorder microphone will always be necessary.

Video Tripods

A video tripod is essential for professional shooting, but even hobby videographers benefit greatly from a good video tripod. Video tripods come in all shapes and sizes, and can cost anywhere from $10 to $1000. With this video tripod list I've tried to find something for everyone. So whether you're looking for something to hold your cell phone steady, or a video tripod that can handle a 15 pound HD camera, there's a video tripod here for you.

Video Bags

A good video bag will protect your gear, organize your accessories, and be easy and fun to carry around. If you're looking for a video bag, this buyer's guide will introduce you to the variety that's available. It includes some of our favorite video bags and carrying cases, and we've tried to include something for every level user and situation.

VIDEO EDITING EQUIPMENT

Get set up to edit videos! Learn about the hardware and software you'll need to get started editing.

Video Editing Computer

Choosing a video editing computer can be tricky. Many old computers won't support video editing at all, and many new computers will only work with the most basic editing software. If you plan to use your new computer for video editing, read this guide to make sure you purchase an appropriate video editing computer system.

Storage Space on the Video Editing Computer

Digital video footage - especially high-definition footage - takes up a lot of drive space, and you'll need somewhere to put it. An external hard drive is one way to solve that problem. But if you buy a video editing computer with a lot of internal drive space, you can put off buying the external hard drive for a while.

Video Editing Computer Inputs

Look at the inputs on any video editing computer you plan to buy. For the fastest way to edit video, the computer should have a firewire input. These inputs are also called IEEE 1394 and iLink. You will use this port to connect your video camcorder to the computer. Or, you can buy an external hard drive with a firewire input and output for storing video footage. You can connect the drive to your computer, and connect the camcorder to the drive.

A USB 2.0 port will work in place of firewire. These aren't as fast, though, and don't give you as many options for connecting external devices to your computer.

Your Plans for the Video Editing Computer

Before you buy a video editing computer, consider the projects that you plan to create. If you only plan on editing basic videos using free software like Movie Maker or iMovie, most new computers out there have the right inputs and plenty of storage space to fit your needs.

External Hard Drive

Video files are BIG (about 1 GB per 5 minutes), so you may need to purchase extra hard drive storage before you can start editing. If you do need more space, you can buy an internal drive to add to your tower, or an external drive. Consider the issues below to determine what is the right solution for you.

Internal or External?

Adding an internal drive to your desktop computer is an easy solution if you have space to put it in your tower. Most electronics stores offer a wide selection of internal drives and have professional service departments to install them. External drives, however, offer a lot of convenience. You can connect them to any computer to watch, edit and share your videos wherever you are. You can also continually buy more, and connect them to each other, so your computer never runs low on storage space.

Speed

The speed of the drive affects how quickly the files it contains can be accessed and transferred. Any hard drive you buy for video should be at least 7200 RPM. Otherwise it will just be too slow for your purposes.

Connection

Generally, an external drive connects to your computer through a USB cable or FireWire. Make sure the connection method is compatible with your computer; that you have a free port to which to connect it; and that you have the appropriate cable (most often, these don't come with the hard drive).

Size

The size drive you buy depends on what you intend to do with it. While you'll generally get better value buying a larger drive, these do have their drawbacks. Hard drives fail—expect to replace them every few years—and thus the larger the drive is, the more work you'll stand to lose if it crashes.

It's preferable to buy a few smaller drives so that you can back up your work and save finished projects separate from those you're still working on.

Portability

Like any other sophisticated electronics, hard drives can be very sensitive, and most aren't designed to be moved around much. If you plan to use yours in more than one location, be sure to purchase a hard drive that is shock resistant. If yours isn't shock resistant, save the original packaging to use when transporting the drive.

Compatability

Most hard drives come unformatted, and can thus work with both Macs and PCs. Make sure that this is the case, though, before you make your purchase.

Warranty

A good warranty is valuable insurance. Spend the time you need to fully understand the terms and conditions of the warranty that comes with your drive. Be sure to fill out and return the product registration card. Always keep a copy of the registration, the warranty terms and your sales receipt with your important records. If your drive crashes before its time, nothing can replace lost video material, but a return of your financial investment may prevent a total catastrophe.

Before You Buy Video Editing Software

Editing software comes in all flavors, from free, online editing software you can use anywhere, to editing software that costs thousands and requires a powerful computer. Which editing software is right for you? Learn about the different types of editing software available.

Try It for Free

Before you buy any video editing software, give the free stuff a try; you may find that it works for your project. iMovie (Macs) or Movie Maker (PCs) comes installed on new computers. If you don't already have one of these video editing programs, you can easily get it for cheap or free. The features, graphics and special effects are often perfect for hobby video enthusiasts as well as beginning video editors looking to experiment. If you'd like to do more advanced editing, you can purchase a new program or customize the one you already have.

Download a Deal

The web is filled with downloads that let you turbo charge iMovie and Movie Maker by adding near-professional audio, visual and graphic effects. Use these add-ons to customize your video editing system based on the features your need.

Try Before You Buy

Video editing software that is more sophisticated than iMovie and Movie Maker generally represents a significant investment. Programs such as Avid, Final Cut Pro and Adobe can cost more than a thousand dollars. Like any purchase this large, you'll want to give it a test run before committing. Local cable access stations are an excellent resource. Many offer free training and equipment use to community members, allowing you to get your hands on high-end editing systems. Schools, libraries and video professionals may also have editing equipment available for you to use or rent.

Find Support

Technical support is crucial to video editing success! Even the most experienced editor encounters problems that aren't covered in the manual. When disaster strikes you'll need a place to turn. Before making a purchase, find out what kind of telephone and online support the software manufacturer offers. User forums and blogs are also useful resources when you experience difficulty—it's likely that someone has asked about the same problem before. Look online for active, informative support groups before you buy, and you'll know where to go when you have a problem later on.

Anything Extra?

Many editing programs come bundled with other software for compressing video, creatingDVDs or other tasks. These add-ons increase the value of the software. They can also ensure ease and compatibility when it comes to performing post-editing tasks.

Free Video Editing Software

Free video editing software is an easy and convenient way to edit your videos, and is great for beginners. Most of the free video editing programs have limited editing features, so after a while you may want to look at the mid-level digital video software or the top professional video editing programs.

- Movie: iMovie comes free with new Macs. iMovie offers many options for editing video and audio, and adding photos, music and narration to your movies.
- Movie Maker: Movie Maker is Window's free video editing software, which comes installed on new PCs. Use the program to create and share high-quality movies.
- YouTube: YouTube's video editor is an easy, free way to trim clips or remix videos. The editing power is limited, though, and you'll never get an uncompressed version of the final product.
- Pinnacle Spin: Video Spin, for PCs, is free video editing software from Pinnacle. Video Spin is specifically designed to take the complexity out of video editing, but it still includes tools, transitions and effects so you can easily jazz up your videos.
- Virtual Dub: Virtual Dub offers simple editing capabilities. You can't get too creative with the software, but you can use it to make simple, clean edits in your video footage. Virtual Dub works with PCs only.
- Wax: Wax is a bit dated-circa 2004-and the interface looks ancient, but the software comes with a very helpful guide and allows for a lot of creativity.
- Zwei-Stein: This free editing software works with Macs, PCs and and Linux systems. It's easy to download, with a colorful, sparse interface.
- Jahshaka: "Powering the New Hollywood," Jahshaka is a free, open source video editing software, that will give you powerful editing capabilities.
- HyperEngine-AV: This is a popular free software for Macs. The "trackless" editing system makes it easy to combine video, audio and photos.
- Free Online Editing Software: Besides YouTube, there are several other web applications that provide free video editing online. These services are great for re-editing and re-mixing web videos, and some even offer to produce DVDs of your edited videos.

Intermediate Video Editing Software

Digital video software can range from free to several thousand dollars. The digital video software programs included in this list retail for no more than a couple hundred dollars and are great for video hobbyists looking to expand their skills.

- Adobe Video Software: Adobe Premiere Elements is the intermediate software in the Adobe Premiere line. It is a powerful and easy to use, yet stripped down, version on Premiere Pro, and even includes Photoshop elements, a basic version of the photo editing software.

- Corel VideoStudio: Corel VideoStudiois a lot of software for a little money. Designed for PCs, Corel VideoStudio comes in three versions. Each offers multi-track HD editing, professional templates, and a wide range of options for exporting edited videos.
- Nero Software: Nero software is an easy and affordable way to edit videos. Nero software includes basic editing tools, transitions, effects and the ability to export movies in a variety of formats, including to DVD.
- Pinnacle Digital Video Software: Pinnacle is the consumer video editing sotware line of Avid, which is the top-of-the-line in professional video editing software. You can take Pinnacle for a spin, so to speak, by trying out the free version - Pinnacle Spin. If you like the feel of that, consider upgrading to one of the Pinnacle Studio digital video programs.
- Sony Vegas Digital Video Software: Sony Vegas offers a line of digital video software, from the basic Vegas Movie Studio HD to the high-end Vegas Pro. All of the programs offer editing, effects and exporting. And, Sony Vegas is quite affordable - even Vegas Pro is less than many other professional grade editing programs.

Professional Video Editing Software

Professional video editing software is a pricey purchase, and you should compare your options before committing. The programs listed here are the best out there for professional video editing. Compare the features, tools, and price point to find the one that's right for you.

Final Cut Pro

Final Cut is the professional video editing program from Apple, and the software of choice for most professional Mac users. It is fast and powerful, with a wide array of video transitions, templates, and effects for enhancing your videos. With Final Cut Studio, you not only get an excellent professional video editing program, but also other software for compression, motion graphics and DVD production.

Avid Media Composer

Avid Media Composer is a powerful SD and HD professional video editor for Mac or Windows. Media Composer is Avid's top video editing program, and it offers integrated DVD authoring, surround sound audio processing and thousands of powerful real-time effects. Media Composer's integrated toolset provides all of the tools needed to create the highest quality productions and finish to tape, DVD or web streams.

Adobe Premiere Pro

Adobe Premiere Pro CS5 software offers a start-to-finish professional video production solution. Using Adobe Premiere Pro you can work with virtually any video format thanks to native editing support for DV, HDV, RED, Sony XDCAM, XDCAM EX, Panasonic P2, and AVCHD. Adobe Premiere Pro is compatible with PCs and Macs, making is a flexible editing solution for professionals.

Sony Vegas Pro

Vegas Pro is the top video editing software from Sony. Designed for professional audio and video production, as well as DVD and Blu-ray Disc authoring, the software includes support for a variety of video formats, numerous video effects and transitions, and even 3D tools.

Video Editing for Macs

Mac video editing software isn't as abundant as that designed for PCs. But thanks to some great offerings - mostly from Apple itself - Mac video editing is fun and easy.

iMovie

Movie is the free video editing software that comes with all new Apple computers. If you have an older version of iMovie, you'll need to pay to upgrade to iMovie 11.

Movie 11 is an intuitive program that lets new and experienced video editors watch video clips, edit video and audio, and share videos on the web or on DVD.

Final Cut

Final Cut Express and Pro are the intermediate and advanced versions of Apple's video editing software. As you would expect, they work seamlessly for Mac video editing.

Online Video Editing Software

Many websites feature online video editing applications that allow you to edit videos online. These online video editing applications are not as feature-rich as editing software that you would install on your computer, but they do allow you to do simple video editing online. And it doesn't matter if you're using a Mac or a PC, as long as you can access the internet

Adobe Premiere Pro

Adobe Premiere Pro software offers a start-to-finish video production solution for Macs and PCs. Using Adobe Premiere Pro you can work with virtually any video format thanks to native editing support for DV, HDV, RED, Sony XDCAM, XDCAM EX, Panasonic P2, and AVCHD.

Video Editing for PCs

There are many types of video editing software for PCs. The following programs are the most popular for PC video editing.

Movie Maker

Movie Maker, the free video editing software that comes with new PCs, is perfect for beginning video editors.

With Windows Movie Maker you can edit and share video and audio files easily on your home PC.

Sony Vegas

Sony Vegas video editing software offers PC users a variety of options. From the inexpensive and easy-to-use Vegas Movie Studio HD to the high-end Vegas Pro, Sony offers a solution for every editor.

All of the programs offer a huge amount of digital effects and output options for your videos.

Corel VideoStudio

Corel VideoStudio is a PC-based editing program that provides user-friendly editing, as well as a wide variety of titles, transitions and templates. Corel VideoStudio includes a wide variety of export options, as well as DVD and Blu-Ray authoring tools.

Pinnacle Video Editing Software

Pinnacle, a division of Avid, makes a variety of PC-based video editing software. You can use Pinnacle Studio to upload edited movies to YouTube, or export for your cell phone.

Adobe Video Editing Software

Adobe video software is some of the best available, and there's a version for every user level and budget, whether you're a Mac or a PC.

VIDEO PROJECTS

Instructions for completing some of the most popular types of video projects.

Video Blogging

A video blog, or vlog, is a collection of videos--your own or your favorites--posted on a web site. It's easy to create a free video blog, and it's a great way to reach out to audiences and show off your work. With a camcorder, editing software and a high-speed internet connection, you're well on your way to producing a successful video blog!

- First, you'll need a reason, or a theme, for your video blog. Video blogs can have many different applications, and can be created for various personal and professional reasons.
- Find a web host for your video blog. There are many free web services that will host your video blog and simplify the vlogging process. I generally recommend YouTube or Vimeo for hosting videos. You can then use the channel page on the video site as your vlog, or you can create a separate blog site - through wordpress or another service - to embed your videos.
- Name your video blog. The content is more important than the name, but having a catchy yet easy-to-remember title can attract more viewers and keep them coming back.
- Prepare some content. You'll probably be able to generate a lot of interest from your audience at first, but if you don't add new content regularly people will lose interest and stop checking in. So, before you go live with your video blog, make sure you have enough content to keep it going for a little while.
- Post your videos. Now you can upload your videos to the web and post them to your video blog. You can improve the visibility of your videos by adding good titles, tags and description.
- Curate videos. Video curation is the easiest way to put together a great video blog. All you have to do is find videos that other people have produced that will be of interest to your audience, and post them on your vlog.
- Promote your video blog. To gain an audience for your vlog, you'll need to let the world know that it exists! You can do this through blog aggregation sites, search engine optimization and good old fashioned self-promotion (i.e. sending out an email to all your contacts).
- Make money from your video blog. Depending on the content, and where you choose to host your video blog, you may be able to make some money from your videos.
- Keep it up! To have a successful video blog you'll need to continuously create and update your content. If you're creating videos about a topic that you love, keeping an up-to-date video blog should be a joy and not a chore.

Viral Videos

Making a viral video is a little about talent and a lot about luck. While there's no surefire formula for making a viral video, these tips will help increase the chances that your next video goes viral.

Making a Viral Video, Tip 1

Put your viral video everywhere. You should post it on every video sharing site you can find. Services like TubeMogul make this easier by letting you upload your video once and then distributing it for you.

Making a Viral Video, Tip 2:

Promote your viral video. Email the link to your mom, your cousin, your high school class. Make it easy for everyone to then forward the video to all of their friends.

Making a Viral Video, Tip 3:

Keep your viral video short. You want this video to be watched as many times as possible, and that's more likely to happen if people can watch the video quickly. Thirty to sixty seconds is generally a good length for a viral video.

Making a Viral Video, Tip 4:

Make your viral video funny. Humor can be hard to get right, sometimes, but it's the surest way to make people want to share your video. Think about all of the viral videos that you've seen—most of them have been comedies.

Making a Viral Video, Tip 5:

Don't make your viral video an advertisement. Some companies have successfully used viral video to promote a product, but "going viral" is usually a spontaneous byproduct of a great video, not an intentional goal.

Making a Viral Video, Tip 6:

Content beats quality in a viral video. Most successful viral videos are made by amateurs, so good content is more important than high production quality. If viewers enjoy what they see, they'll overlook a shaky camera or focus troubles.

Making a Viral Video, Tip 7:

Don't use copyrighted content in your viral video. If you hope to have your viral video seen by thousands of people, make sure you're not using any unlicensed music or images that could get you in trouble with record companies, television networks and their lawyers.

Making a Viral Video, Tip 8:

Keep trying to make viral videos. Often, it's completely random videos that go viral, and no one can predict with much accuracy what is going to catch on and become a big hit.

Video Interviews

Video interviews, or "talking heads", are common in all types of videos, from documentariesand newscasts to marketing videos and customer testimonials.

Producing a video interview is a straightforward process that you can complete with nearly any type of home video equipment.

Difficulty: Easy

Time Required: Varies

Here's How:

- Prepare yourself and your subject for the video interview by talking about the information that you're going to cover and the questions that you're going to ask. Your subject will be more relaxed and the video interview will go more smoothly if you've talked it out ahead of time.
- Find a good backdrop for conducting the video interview. Ideally, you'll have a location that illustrates something about the person you are interviewing, such as their home or workplace. Make sure that the background is attractive and not too cluttered.
  If you can't find a suitable backdrop for the video interview, you can always seat your subject in front of a blank wall.
- Depending on the location of your video interview, you may want to set up some lights. A basic three-point lighting setup can really enhance the look of your video interview.
  If you're working without a light kit, just use whatever lamps are available to adjust the lighting. Make sure that your subject's face is brightly lit, without any odd shadows.
- Set up your video camera on a tripod at eye-level with your interview subject. The camera should only be three or four feet from the subject. That way, the interview will be more like a conversation and less like an interrogation.
- Use the camera's eyepiece or viewfinder to check the exposure and lighting of the scene. Practice framing your subject in a wide shot, medium shot and close up, and make sure that everything in the frame looks right.
- Ideally, you'll have a wireless lavaliere microphone for recording the video interview. Clip the mic to the subject's shirt so that it's out of the way but provides clear audio.

A lavaliere microphone will not get a good recording of you asking the interview questions. Use another lav mic for yourself, or a microphone attached to the camera, if you want the interview questions recorded as well as the answers.

If you don't have a lav mic, you can always use the camcorder's built in microphone for the video interview. Just make sure the interview is done in a quiet space and that your subject speaks loudly and clearly.

- Seat yourself right next to the camcorder on the side with the flip-out screen. This way, you can subtly monitor the video recording without directing your attention away from the video interview subject.
  Instruct your interview subject to look at you, and not directly into the camera. This will give your interview a more natural look, with the subject looking slightly off camera.
- Press record and start asking your video interview questions. Make sure to give your subject plenty of time to think about and frame their answers; don't just jump in with another question at the first pause in conversation.
  As the interviewer, you need to be completely quiet while your interview subject is answering questions. You can respond with support and empathy by nodding or smiling, but any verbal responses will make editing the interview very difficult.
- Change up the framing between questions, so that you have a variety of wide, medium and close up shots. This will make it easier to edit different segments of the interview together, while avoiding awkward jump cuts.
- When you finish the video interview, leave the camera rolling for a few extra minutes. I've found that people relax when it's all over and start talking more comfortably than they did during the interview. These moments can yield great soundbites.
- How you edit the video interview depends on its purpose. If it's purely archival, you can just transfer the whole tape to DVD without editing. Or, you may want to watch the footage and choose the best stories and soundbites. You can put these together in any order, with or without narration, and add b-roll or transitions to cover any jump cuts.

Marketing & Promotional Videos

Promotional web videos are an exciting internet marketing tool for businesses large and small. Like a traditional TV commercial, a promotional web video will advertise your business's advantages to potential customers. Unlike traditional TV commercials, promotional web videos can be broadcast directly to your target audience for free through email, search engine marketing and video sharing sites such as YouTube.

Producing a promotional web video doesn't have to be complex or expensive. With a little bit of planning, you can produce a promotional web video that becomes a priceless marketing tool for your business.

Difficulty: Average
Time Required: Varies
Here's How:

- Identify Your Goals for Your Promotional Web Video: Before you begin to produce your promotional web video, you need to think about what you want the video to accomplish. Some questions to consider include:
  - Who is the target audience for your promotional web video?
  - What will be the tone of your promotional web video? Funny? Professional? Sincere?
  - What do you want viewers to do after watching your promotional web video? Email it to a friend? Call your company? Click for more information?
- Set a Budget for Producing Your Promotional Web Video: Producing a promotional web video is much less expensive than producing a traditional television commercial. If you're video-savvy and not looking for a super-polished video, you may be able to produce your promotional video yourself for little or no cost.

To produce a high-quality promotional web video, though, you may want to consult a professional web video production company for assistance. Many companies work with businesses to develop, produce and distribute promotional web videos.

- Plan Your Promotional Web Video: In the pre-production phase of your promotional web video, you'll need to plan out how the video will look and sound. This process can be very formal, including scriptwriting and mapping out each frame of the video.

Or, if you want a documentary-style promotional web video, you can be less formal. Think about what themes you want to address, what footage you want to capture, and who will act as a spokesperson in the promotional web video. If you're working with a professional production company they can help you with the planning a scriptwriting process.

- Shoot Your Promotional Web Video: If you've developed a good plan, shooting your promotional web video should go smoothly. By knowing exactly what footage you need, you'll save a lot of time and, if you're working with a professional video production company, money.
- Edit Your Promotional Web Video: Again, with a good plan editing your promotional web video should be a breeze. If you're doing it yourself, our video editing tutorials can help you with adding titles, music and images to your promotional web video.
- Post Your Promotional Video to the Web: There are many places on the web where you can post your promotional web video. The

first and most obvious is on your web site. If you are working with a professional production company, they can post the video on your homepage or even design a web page specifically for displaying the promotional web video. If you are producing the video on your own, it may be easier to just post the video on YouTube and then embed the YouTube video on your web site.

- Share Your Promotional Web Video: Once your promotional video is posted on the web, you'll need want to have it seen by as many people as possible. There are many ways to gain an audience for your video, including:
- Posting your video on a variety of video sharing web sites
- Publicizing your video on iTunes
- Emailing your video to friends and colleagues

School Videos

University marketing videos are an effective way to promote your school online to prospective students, as well as communicate with alumni.

Time Required: Depends

Here's How:

- Plan your university marketing videos. University marketing videos can be anything from short tv commercials to mini documentaries profiling students and teachers at your school. If you need creative help, we've got lots of ideas for school marketing videos.
- Get a crew and equipment for your university marketing videos. If you want a professional university marketing video, consider hiring a professional video production company.

If you don't have the budget for a professional production, talk to your university media center to see if they can provide you with equipment and people to operate it. If your university has a media production program, you may be able to find students who have the skills to produce your university marketing video.

- Find students and teachers to participate in your university marketing video. Using real students and teachers will give your university marketing video an authentic feel. Instead of giving them scripted lines to recite, come up with good interview questions that will lead your subjects to saying what you want in their own words and their own way.
- Location scout your university marketing video. What buildings and campus locations do you want to show off in your university marketing video? When choosing locations, think about when you'll be shooting, whether there will be people around, and what the sound and light will be like.
- Record your university marketing video. Once everything is

planned, you're ready to begin recording your university marketing video. It's vital that you get high-quality, professional looking footage, so make sure to shoot and re-shoot until you get everything right. You may end up recording as much as 1 hour of footage for each edited minute in the final video.

- Distribute your university marketing video. You can always put your university marketing video on DVD and mail it to any interested student. This can get pricey, though. Alternatively, distribute the video for free on your university web site and video sharing web sites such as YouTube.

Webcam Videos

Recording webcam footage is an easy and convenient way to make videos. However, many webcam recordings suffer from very poor video and audio quality. But by making a few adjustments you record with your webcam, you can quickly and drastically improve the quality of your webcam videos.

Difficulty: Easy

Time Required: 15 minutes

Here's How:

- Set up your webcam. Many computers now come with a webcam built in at the top of the screen. Otherwise, you can purchase an inexpensive webcam that connects to your computer via USB. It's also possible to use many digital camcorders as webcams, by connecting them to the computer and recording directly onto the hard drive.
- Set up your audio recording. Most computers have built-in microphones, but you'll get better sound quality using an external microphone. Even a basic desktop microphone that connects to your computer will greatly improve the audio quality of your webcam recording.
- Position your webcam. Too many webcam recordings feature people staring at their computer screen, while the webcam stares down at them from above. Avoid this by placing your webcam at eye-level. When recording, make sure to look into the webcam and not at the recording screen.
- Clean up the background. I know that many webcam videos are recorded in people's homes or bedrooms, and the intimacy of the setting is part of the charm. However, it's important to make sure that the background isn't to cluttered, and that there's nothing inappropriate showing on the screen.
- Adjust the lighting. The brighter it is, the better your webcam recording will look. For a simple solution, you can use your household lamps to approximate three-point lighting.

- Record, record, record. Once you have your equipment set up, you're ready to start recording with your webcam. The nice thing about webcam video is that it's easy to do multiple takes until you get something perfect, but viewers aren't looking for perfection, so you can easily get away with quick takes and off-the-cuff videos.

How to Videos

You can make how to videos to show off your skills, explain a product, or promote your business. If you have basic videography and editing equipment, it's not too difficult to make how to videos.

Difficulty: Average

Time Required: Depends

Here's How:

- What will you make how to videos about? The first thing you'll need to do, before you start to make how to videos, is to plan them out. What's the topic, who's the star, what props will you need? You don't necessarily need to script the entire video, but it will help if you have a basic outline and know what points you'll emphasize before you start to make how to videos.
- Where will you make how to videos? You'll need to find a good location to make how to videos. Pick a quiet spot with good lighting (or plenty of space to set up extra video lights) and a nice background. If you want to make how to videos about cooking, find a nice kitchen set up. If you're going make how to videos for your business, you might want to make sure your logo is visible in the background.
- Do you have the right equipment to make how to videos? Under the right circumstances, you can make how to videos that look great with nothing more than a Flip camcorder. However, to make professional looking how to videos you'll want, in addition to a camcorder, a lapel microphone for the video star to wear, a tripod to hold the camera steady and video lights.
- Videotape the how to presentation. When I make how to videos I always make sure to get plenty of different takes and angles. Usually I'll tape one run through of the demonstration as a wide shot, so I can see the subject talking and see what they're doing. Then, I ask them to give the presentation again, and this time I get lots of close ups. I'm also sure to get individual shots of all the products and materials used in the how to demonstration.
- Edit the how to video. If you did a good job shooting the how to video, editing should be a snap! When I edit how to videos, I start by listening to the audio. I cut out any mistakes or anything extraneous and put the remaining sound bites together to get the

audio portion of the how to video. Once the audio is done, I add the close up shots to the video in places where I need them, either to cover up a cut or to provide a good visual explanation. Lastly, I add title graphics where necessary and music to enhance the video.

- Publish your how to videos. Once you've finished making the how to videos, you can share them online. There are many how to video websites that accept user submissions. You can upload to other video sharing sites like YouTube, or put the videos on your own web site.

Wedding Videos

Summers are wedding time, and weddings mean wedding videography. If you're planning to videotape weddings these tips will show you how to shoot wedding videos that look great.

Remember Your Role

When you're videotaping a wedding you're generally either doing it as a friend or professional who's been entrusted to shoot the official wedding video, or as a guest who happened to bring along a camcorder. If you're not shooting the official wedding video stay out of the way of the person who is. The bride and groom likely paid a lot of money to hire this professional, and he or she should always be given priority in setting up the best shot and getting the best angle of the events. If you step in front of the hired videographer to get a good shot of the vows, you're actually ruining the wedding video that the bride and groom paid for. No one will be happy with you, no matter how good your video looks.

Be Prepared

If you're new to videography, shooting wedding videos makes for an intense boot camp. The tips for recording good video and good audio will help with shooting a wedding video (or any type of video for that matter).

Tapes & Batteries

You'll need five or six tapes, depending on length of the day. Label the tapes ahead of time so that it's easier to keep them in order. You'll also need an extra battery or two, as just one probably won't last you through the whole day. If you don't have enough batteries be sure to bring your charger so that you can recharge the batteries during down time.

Use a Lapel Mic

Without a lapel microphone for the groom you probably won't be able to hear the audio for the vows. Ideally, you'll have a wireless microphone which can hook into your camera. However, these are expensive, so you may not be able to afford one (especially if you're not getting paid for your work!). As an alternative, you can buy a digital recorder (or transform your iPod into a digital recorder) and wire a lapel mic into that. You'll have to synch the audio and video while editing.

Know the Schedule

Talk to the couple ahead of time to find out the schedule for the wedding. That way you'll be able to anticipate the action and won't find yourself running out of tape at an crucial moment or missing an important event that you should be videotaping. Ideally you'll be able to attend the wedding rehearsal. This will give you a chance to find the best place to set up your camera. You'll also have an opportunity to find out if there are any restrictions at the ceremony site. Many churches have rules about where videographers can stand, whether you can move around, and about the use of lights. If you weren't at the rehearsal grab a copy of the program so that you can figure out what's going to be happening during the ceremony.

Be Unobtrusive

Remember, a wedding is a day to celebrate the couple that's getting married. While it's important that you make a great video to remember that day, it's just as important that you let the bride and groom and their guests enjoy the day. You may have to move around some during the ceremony but try to do it quickly and quietly so as to not draw attention away from the couple. Also, use your zoom to get close ups of the guests. No one likes to have a camera shoved in their face, and it's one of the biggest complaints people have about wedding videographers.

Talk to the Guests (or Leave Them Alone)

Some wedding guests are vocal and want to say something to the camera. Some are camera shy and want to be left alone, if that's the case, respect their wishes.

Light the Scene

Thanks to new, better quality digital camcorders, gone are the days when wedding videographers needed to set up large, 1000 watt lights. Still, though, you may need some extra light to get good footage during a wedding. A small, 50-watt light mounted on the top of your camera will light the scene without blinding guests or breaking your budget.

Make Friends With Other Vendors

The videographer, the dj, the photographer and the reception site coordinator all have a common goal: make the day go smoothly for the bride and groom. As soon as possible introduce yourself to these people and find out what you can do to work together to all do your jobs well. The photographer should know where your camcorder will be set up at the ceremony, so he or she doesn't stand in front of it. The dj or site coordinator can tell you the schedule of events for the reception, and make sure that you're in the room whenever anything important happens.

Take a Break

Shooting a wedding video means spending a long day on your feet and hard at work. Be sure to take a break now and then for some rest and refreshment. I don't recommend drinking on the job, but a Coke or an ice water can revive your spirits when you start to fade. Also, taking a break can be good for guests who are camera shy. Some people will leave the dance floor the instant they see a video camera coming their way. If you take a break and sit out a few songs, you'll give these folks a chance to have some fun without fear or embarrassment of their dance moves being caught on tape.

Try Two Cameras

If you have two video cameras use both of them to shoot the wedding video. That way you can set one up to capture a wide shot of the bride, groom and officiant, and use the other one to get close ups and reaction shots. By using two cameras you know you'll always have the wide shot to cut away to, which will give you more flexibility during editing and shooting.

Get the Shots

Every wedding is unique, but there are certain things that are common to most weddings. Thiswedding videography checklist should help you make sure you get the important shots that the bride and groom will expect to see in their wedding video.

Graduation Videos

Graduation videos are easy if you follow a few basic steps. Keep these tips in mind for your graduation video. It'll be fun to watch and a great way to remember the day.

- Polish Your Skills: If you're new to videography, make sure that you know the basic steps to shooting great video. These videotaping tips will help you get high-quality video shots during any event, and they'll be especially helpful while videotaping graduation ceremonies.
- Battery Power: You'll need to use batteries to power your camcorder, since there will probably be nowhere to plug in. Make sure the batteries are fully charged. You may even want to invest in an extra battery for videotaping graduation ceremonies - it'll be a long day and you want your camcorder to make it all the way through!

- Use a Tripod: Graduation ceremonies last a long time, and you don't want your arms to get tired and your filming to get shaky. A tripod will help you maintain a still, steady shot for the duration.
- Arrive Early: By arriving early, you'll be able to get the lay of the land and find a good place to set up your camcorder. You'll want plenty of time to make sure you get a good view of the stage and the students, and to check that your camcorder is working properly.
- Consider the Sound: You may not have many options when recording audio at the graduation ceremonies. If possible, though, try to set up your tripod near an audio speaker. That should make it easier for your camera to pick up the speaker's voice, instead of background noise from the audience.
- Read the Program: After you've set up your camcorder, grab a program and familiarize yourself with the order of events. By knowing when things will be occuring you'll be able to anticipate the action and have your camcorder ready to capture the important moments.
- Respect Others: Sure, it's important that you do a good job videotaping graduation ceremonies and making your graduation movie the best it can be. But don't forget that there are many others who want to enjoy the event and celebrate the graduates' accomplishments. Be conscious about where you are standing and try to stay out of the way of other people.
- Videotape Personal Remembrances: There's more to graduation day than speeches and handing out diplomas. Videotape interviews with the graduate and his or her friends and family. Ask about the importance of the day, their favorite memories of school and their plans for the future.
- Don't Forget B-Roll: You should always shoot b-roll during any video production. It will add visual interest to your finished video, and give you more options while editing.
- Have Fun! Last but not least, don't forget to celebrate! Graduation is a big day, and you want to enjoy it together with your family, not alone behind the video camera.

Family Video Memoir

Recording your family history on video is a worthwhile project. With the right equipment, a little bit of training and a lot of curiosity you’ll be able to interview relatives, preserve photos and create a video DVD for the generations. Learn how to use Apple iMovie to create a family memoir video.

Difficulty: Easy

Time Required: varies

Here's How:

- Conduct interviews with older family members: To learn about your family history, go to grandparents, aunts and uncles, and other relatives. Ask them about their lives, as well as what they know of family history.

Place the camera on a tripod when you're shooting the interview. If you have a lapel mic for your interview subject, you'll be able to record much clearer sound.

- Set up a new video project on your computer: You don't need a complex editing system. With iMovie you can edit the interviews and add titles and photos to your movie.
- Edit the interviews: You'll probably want to edit your interviews to cut out over-long answers and make the video flow more smoothly. Use fades, dissolves or b-roll to avoid jump cuts in your editing.
- Create a photomontage: Enhance your family memoir DVD by adding a family photo album to the video. Include photos of your distant relatives, if you have them, as well as current pictures of your family today.
- Add titles and credits: Give your family memoir video a professional look by adding a title at the beginning, and credits at the end. Use the credits to thank everyone who helped you to create the video.
- Share the family video memoir: Surely there will be many family members eager to watch your family memoir DVD. You can share the project with them by burning it to DVD or uploading it to the web.

Vacation Videos

If you're taking a vacation, I hope you're bringing your video equipment. With a small camcorder you can make a vacation video that will leave a big impression and big memories down the road. Just follow this vacation video packing list and be sure you have everything you need.

Vacation Video Packing List Item 1: Tapes & Disks

If you use a mini-dv or mini dvd camcorder for your vacation video, make sure that you have plenty of tapes or disks. Sure, you can buy them on the road, but it's easier (and cheaper) to have them all stocked and ready to go. For extra convenience, label each tape or disk ahead of time (2009 Summer Vacation, Tape 1 etc.) This way, you'll be able to keep them in order and you'll know quickly which tapes have been used and which are still blank. Also, once you get home you'll avoid the problem of losing track of the tapes.

Vacation Video Packing List Item 2: Hard Disk Storage

If you're shooting your vacation video with a hard disk camcorder, bring something so that you can offload footage from the camera when the disk gets full. You'll either want some extra memory cards, or a laptop computer and external hard drive to which you can transfer the footage.

Vacation Video Packing List Item 3: Batteries

Make sure that you have enough battery power to get you through the day. You'll be able to recharge the batteries at night, but when you're out on a day-long adventure you may need a second or even a third battery to be able to do all the shooting you want for your vacation video. If you're traveling to a foreign country, make sure you have the proper current adapters so you can plug in the battery chargers.

Vacation Video Packing List Item 4: Microphones

When shooting your vacation video, you'll probably just end up using your camcorder's microphone most of the time. If you have any external microphones, though, it doesn't hurt to bring them along, as they could really enhance your vacation video.

You can ask your tour guide to wear a lapel mic, so that you pick up everything he or she says, or you can give your kids a handheld mic so they can do news broadcasts from the sites you visit.

Vacation Video Packing List Item 5: Cleaning Supplies

A brush and a soft cloth for your lens will be indispensable while you're on the road making your vacation video. Use your cleaning supplies liberally to keep your camcorder clean and clear while you travel.

Vacation Video Packing List Item 6: Camcorder Bag

It's best for your camera to travel incognito. It will be in and out of your car, and carried around from site to site. If the bag is obviously carrying a pricey camcorder, it will be more tempting to thieves. Instead, pack the video camera in an old backpack or some other bag that will be discreet. Also, be sure to label your camcorder bag with a cell phone number or some other phone number that you'll be able to check while you're on your trip. That way, if it's lost it can be recovered more easily.

Vacation Video Packing List Item 7: Other Accessories

If you're really serious about your videography, you might want to consider bringing accessories for your video camera like lenses, filters and a tripod. These video accessories are helpful, but in my experience they mean that you'll be spending more time making your vacation video, and less time relaxing on your vacation.

Remember, vacation videos are about preserving memories. If you spend all your time fussing with the video, that'll be the extent of your memories when you get home.

Video Virtual Tours

Online virtual tours are perfect for Realtors, hotels, restaurants, tourist attractions and anyone who wants to show off a space. It's easy to videotape a virtual tour and post it on your web site.

Difficulty: Easy

Time Required: 1 hour

Here's How:

- Stage the virtual tour area. A few decorative touches can help make your property look its best in the online virtual tour. Make sure that all rooms are clean, and that any clutter is hidden. Adding a few flowers or a picture on the wall can also increase the appeal of a virtual tour.
- Light the virtual tour space. Plenty of lighting improves the look of your virtual tour. Do the videotaping during the day, and turn on as many lights as possible.
- Set up your virtual tour shots. You want to videotape your virtual tour with your camcorder on a tripod in order to get a smooth, steady shot. Set up either in a corner, so you can get a wide view of the room, or in the center so you can do a 360-degree pan.
- Keep the virtual tour slow. When you are shooting the virtual tour keep the camera moving slowly. A smooth, slow shot will allow viewers to really appreciate what they're seeing on the tour. A quick, jittery shot, on the other hand, can leave viewers feeling queasy.
- Edit the virtual tour. Once you have videotaped all of the sections of your virtual tour, it's time to edit your virtual tour. Editing doesn't need to be complicated, and can be done with simple, free video editing software. Just add a title to the beginning, and use simple transitions, like fades, between each of the parts of the virtual tour. To enhance the mood of the virtual tour, add some royalty-free music.
- Post the virtual tour online. Depending on its purpose, you may want to upload your virtual tour to a personal web site, or use one or more of the many video sharing sites out there for free distribution. If you want the virtual tour to show up in search results, be sure to use these video SEO tips.

Top Family Video Projects

Find video projects to work on for or with your family. These family video projects give you the creative inspiration you need to take out the camera and start making movies with your family.

- Events and Celebrations: Birthdays, holidays, school plays, family reunions--these are the reasons you bought a video camera in the first place! Now just remember to use it. You'll be happy to be able to look back on these memories in years to come.

- Family Recipes: A lot of family recipes can't be written down because they only exist in grandma's head. But if you get her to make her famous cookies in front of the camera, you'll always be able to replicate them.
- Interviews: Everyone has stories to tell, they're just waiting to be asked. Ask mom and dad about being kids; ask the kids about growing up; or ask grandma and grandpa about the good old days.
- Photomontages: Bring your photos to life by moving your picture albums to DVD. Add movement and music to old or new photos so they're even more fun to share.
- Transfer Old Family Videos to DVD: You'll watch your old family videos a lot more often if you convert them from VHS to DVD. Thistutorial will show you an easy way to complete the project.
- Video Blogs: A video blog lets you share any family video projects with friends and relatives around the world.
- Give the Kids the Camera: If the kids are old enough, let them use the video camera for the day. They'll have a lot of fun, and you'll get to see the world from their perspective.
- Yearbook Videos: Package a year's worth of videos together in one "yearbook" to commemorate the passage of time. These can make great birthday presents, or end-of-year projects to complete as a family over winter break.
- Make a Movie Together: Working together as a family to put together a movie can be a rewarding project. You can re-create a scene from your favorite movie, adapt a story, or write one yourselves. Everyone can act in the movie and be part of the production crew; it'll make for a fun day together, and you'll create a lasting memento.
- Home Tour: Take the camera on a room to room tour of the house. This is something your insurance adjuster will love you for, and it's a nice keepsake if you're moving. You can also record narration over the tour with family members talking about their memories and favorite things about your home.

Index

I

M

N

O

P

R

S

T

V

W